Light

HOW TO SEE IT
HOW TO PAINT IT

The Parade, Weymouth by Arthur Maderson, oil on canvas, 28in×19in (70cm×48cm)

Light

HOW TO SEE IT
HOW TO PAINT IT

Lucy Willis

edited by Angela Gair

STUDIO
VISTA

STUDIO VISTA

A Cassell Imprint

Cassell
Wellington House
125 Strand
London WC2R 0BB

First published 1988
Reprinted 1989, 1990
First paperback edition 1991
Reprinted 1993
Second paperback edition 1997

British Library Cataloguing in Publication Data
A catalogue record
for this book is
available from the
British Library

ISBN 0-289-80170-2

This book was designed and produced by
Quarto Publishing plc
The Old Brewery, 6 Blundell Street
London N7 9BH

Project Co-ordinator: Angela Gair
Senior Editor: Maria Pal
Designer: Bob Gordon

Art Director: Moira Clinch
Editorial Director: Carolyn King

Typeset by Text Filmsetters Ltd and
Burbeck Associates Ltd
Manufactured in Hong Kong by
Regent Publishing Services Ltd
Printed by Leefung-Asco Printers Ltd, China

CONTENTS

INTRODUCTION

For those who have never tried to paint, the artist's preoccupation with light may seem curious. Is it really necessary to worry about seeking out 'good' light? Can the light on a rainy day be as interesting as that of a sunny day? Is there ever any reason for us to say "I can't paint, the light's all wrong"? How does light affect the mood and content of a painting? What is the relationship between light and colour? How on earth do you paint something so insubstantial as light, and why is it so important?

In this book I hope to demonstrate not only how light reveals, models and illuminates the world, but how it can also be a great source of inspiration to you the painter. There's no need to pack your paints and fly off in search of dazzling Mediterranean sunshine; light is all around you. Once you become *aware* of light and its myriad effects, your most familiar surroundings, no matter how mundane they may once have seemed, become amazingly rich and beautiful. Everyday objects – your breakfast table, the coffeepot, the cup you drink from – are miraculously transformed as you begin to see them in terms of subtle variations of color and tone.

When I was an art student I used to spend a lot of time agonizing about what to paint next. I would try to find subjects that were in some way 'important'; I was constantly in search of something that would really inspire me. At the end of the day, however, despite all my struggles, I rarely ended up with a picture that was inspired at all. It began to dawn on me that the painters who impressed me most were those who somehow captured a feeling of light in their work. My own paintings began to reflect this concern. I suppose that this was largely an unconscious development, and it wasn't until I began to teach people to draw and paint that I realized how crucial it is to have a sound understanding and appreciation of the way light works. My old problem of finding a subject had solved itself: everything, no matter how simple or ordinary, could now become a source of inspiration.

Essentially, painting is in the first place a matter of seeing: we must develop the ability not only to respond to and enjoy the visual world around us, but also to analyze and understand it. What is it that enables you to do this? What special skills do you need, as a painter, to make sense of the way in which light affects and determines what you see? How do you set about capturing these effects in paint? In this book I hope you will find some of the answers. The following chapters will help you discover how to:

◆ Evaluate light in terms of tonal value
◆ Use colour to create dazzling impressions of light
◆ Paint shadows and reflected light in order to create depth, form and luminosity
◆ Tackle the problems of recording the transient effects of outdoor light
◆ Control the lighting to the best effect when painting indoors

As well as a discussion of all these aspects of painting, there are practical projects designed to help and encourage you to make some explorations of your own. There are also many paintings, by a wide range of artists, specially chosen to illustrate the points that are made, to demonstrate the techniques covered, and to give you ideas and inspiration. You will see how various artists have handled the painting of light in a great diversity of styles, which should help to clarify your own work. It is, above all, the impression of light which brings these paintings to life and communicates a sense of the artist's enthusiasm and enjoyment. As your insight and skills develop you will begin to see these qualities emerge in your own pictures, filling them with life, colour and light.

James McNeill Whistler, whose paintings so beautifully interpreted the effects of light, said: "As music is the poetry of sound, so painting is the poetry of sight." With this in mind, paint on!

The Grapefruit Grove by Lucy Willis, watercolour, 15in×11in (37cm×27cm)

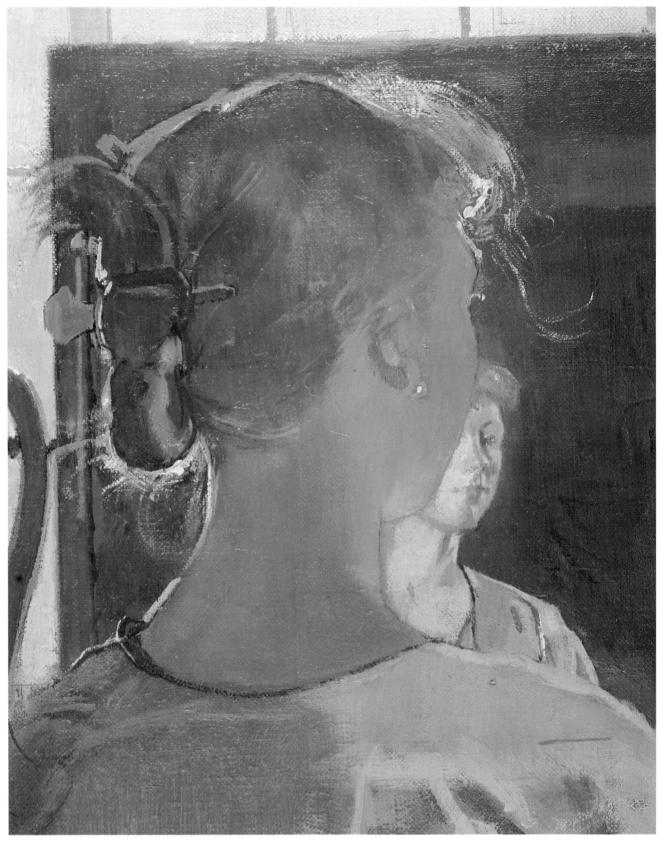

CHAPTER ONE

THE POETRY OF LIGHT

Any painting which aims in some way to represent the world around us by creating the illusion of reality, is a painting about light. None of the considerations with which the painter is occupied – tone, colour, form, space – is independent of it.

Light can be the source of inspiration as well as the means of revelation. When light sparkles on water, catches and animates a face, shines through the leaves of a tree, or gently illuminates a misty morning, we, as artists, are captivated and inspired. Yet light itself is an elusive and transitory thing. How can we even begin to capture its poetry on paper or canvas?

On the following pages you will discover some of the ways in which different artists – some great masters of the past, some painters working today – have captured and interpreted the effects of light. I hope you will be guided by their inspiration, just as each of them was guided, in turn, by the inspiration of artists who went before them.

MANY ARTISTS have a certain favourite theme or motif which appears frequently in their paintings. For the English artist Joseph Mallord William Turner (1775–1851), that theme was light. And it was the magical effects of the light of Venice, that "city of shimmering spaces", which had the most profound effect on Turner's work. *Venice: Punta della Salute* is one of only four watercolour wash studies that he made on his first visit to Venice in 1819, in which he recorded the transient effects of light on sky and water. Turner used his watercolours with breathtaking economy,

△ **Venice: Punta della Salute** by JMW Turner, 1819.
Watercolour, 9in×11¾in (23cm×30cm).

applying transparent washes of pure, vibrant colour which allow the white paper to shine through luminously. Note how the buildings in the distance are reduced to a single wash, creating a pale silhouette, and how warm pinks and yellows bring the foreground closer, increasing the sense of space. Despite its brevity, this refreshingly simple study has an almost majestic feeling of air and light.

JOHN NEWBERRY is a contemporary painter whose watercolours, like those of Turner, are concerned above all with light and atmosphere. In this small painting of a north African town, *Bab Zuwayla, Cairo*, Newberry has used the subtle tonal qualities of watercolour to capture the soft radiance of early evening light. The sky is laid in with a thin, transparent wash of blue, graded almost to nothing as it nears the horizon, and gaining luminosity from the white paper beneath. The distant buildings, backlit by the evening sky, dissolve into pale, shadowy forms, which Newberry renders with simple blocks of tone together with the

△**Bab Zuwayla, Cairo** by John Newberry, watercolour, 7in×8in (17cm×20cm).

merest suggestion of detail. The limited palette of soft, warm neutrals lends an overall harmony which enhances the tranquil mood of the scene.

Although the subject and the time of day portrayed in this painting are quite different from those of the Turner opposite, there are nevertheless certain similarities in the handling of the paint and in the artist's observation of the way light suffuses everything with its radiance.

OF ALL THE GREAT Impressionist painters, Claude Monet (1840–1926) was the one most concerned with the study of light and its effects on natural forms. In his later years, light became his sole obsession, and he maintained that his only aim was "to paint directly from nature, striving to render my impression in the face of the most fugitive effects." *Grain Stacks, End of Summer*, one of Monet's most beautiful and best-known paintings, was one of a series which he painted in a field beside his house at Giverney in the summer and autumn of 1890. His preoccupation in this painting, as throughout the series, was not so much with the subject matter itself as with the atmosphere and the light; notice, for example, how the physical bulk of the haystacks appears to be dissolved in light. As the day prog-

△**Grain Stacks, End of Summer** by Claude Monet, 1890. Oil on canvas, 23¾in × 39¼in (60cm × 100cm).

ressed, each change in the colour of the light provoked a new canvas, and so the series developed. Monet would work each day in rotation on the paintings, changing canvases as the conditions changed. Each painting is remarkable in itself, and the whole series constitutes one of the most intense investigations into the nature of light in the history of painting. Through his brushwork and his use of colour, Monet achieved a deep and vibrant luminosity which is diffused over the whole canvas.

△**Orchard, Point of Sunset** by Arthur
Maderson, oil on canvas, 32in×24in
(80cm×60cm).

ARTHUR MADERSON is a contemporary painter whose main concern is also with the painting of light. He has adopted some of the same principles as Monet and the other French Impressionists, using broken brushstrokes of thickly applied paint to create deep harmonies and vibrant juxtapositions of colour. Like Monet, Maderson sees as much colour in shadows as in bright sunlight. In *Orchard, Point of Sunset* he intersperses the deep foreground shadows – built up in patches of greens, blues, and purples – with dabs of intense, warm orange. The painting has a homogeneity and unity which is produced by working all parts of the canvas with equal emphasis: the figure in the foreground is painted in the same spontaneous manner as the orchard trees. Maderson ignores superficial detail and concentrates on relationships of tone and color in evoking the striking effects of evening sunlight.

IF MONET WAS the most fanatical of the Impressionists, Pierre Auguste Renoir (1841–1919) was undoubtedly one of the least fanatical painters of that school. To Monet, the play of light was more important than what it played upon; to Renoir, however, people were the important thing, and he delighted in rendering glowing skins and rich, lustrous textures. Renoir also had a traditionalist's feeling for solid forms and structured design, which is evident in this delightful figure composition, *The Luncheon of the Boating Party*. The whole painting is about pleasure, and the artist's own enjoyment of his subject is apparent, not only in the sensuality of the figures, but also in the lovingly painted still

△ **The Luncheon of the Boating Party**
by Pierre Auguste Renoir, 1881. Oil on canvas, 51in×68in (129cm×173cm).

life on the table. Renoir is famous for his descriptions of pearly, luminous flesh, but we can see that he was equally interested in the way that light plays on a white shirt or tablecloth, a glass, or a bottle of wine. Although there are no strong shadows here, a deep luminosity is created by delicate passages of reflected colour and light. The soft, diffused light under the awning adds to the warm, intimate atmosphere of the scene.

JOHN WARD is a contemporary painter well known for his figure compositions and his portraits. *The Beach Party* has many elements in common with Renoir's painting opposite, including the same charmingly informal atmosphere which captures perfectly that "summer vacation" feeling. The composition is similar, too, with the strong verticals of the poles, the horizontal planes of table and awning, and the relaxed disposition of the figures. The light, however, has been handled in a slightly different way, with strong tonal contrasts between the figures under the shade of the

△ **The Beach Party** by John Ward, oil on canvas, 54in×72in (137cm×183cm).

awning and the brightly lit beach beyond. Where patches of light do break into the shadowy foreground, on legs and shoulders for example, they form distinct shapes of contrasting tone. It is this balance between the dark and light areas of the composition which conveys both the sleepy, languid atmosphere under the awning and the intense heat and brilliant sunshine beyond.

Homeward Bound by Moonlight by Albert Goodwin (1845-1932). Watercolour and body colour, 12in×16in (30cm×40cm)

CHAPTER TWO

TONE AND LIGHT

The tonal distribution of light and shade is vital in creating the illusion of reality in a painting. It turns the flat, two-dimensional surface of canvas or paper into a three-dimensional world of depth, space and light. It is difficult for us today to imagine what a momentous point it was, in the early history of painting, when artists discovered how tone helped them to model form and to change what had been flat symbols of the world they knew into solid representations of reality.

In this chapter you will discover how to judge the tonal values of your subject and match them in your painting, so as to capture the beauty and sparkle of nature.

Understanding Tone

When you first begin painting it may strike you that your pictures appear somewhat flat and lifeless. More often than not, the reason for this is that your tonal contrasts are not strong enough. It takes a certain degree of courage to render shadows as dark as they really appear and to keep light passages in a picture bright enough. However, once you become familiar with the idea of tone, and are comfortable with a concept that might require you to paint something you know to be white as almost black, you will be astounded at the improvement it makes to your pictures, which will take on greater strength and realism.

Tone is an often misunderstood word, but it really has a very clear meaning. 'Tone' merely refers to how light or dark a colour appears to be. The tonal 'scale' runs from black to white, and every colour has a tonal equivalent somewhere on that scale. Imagine comparing a colour photograph and a black-and-white photograph: you will notice how pale colours like yellow photograph as light grey and therefore have a high tonal value. A dark red or green might register as dark grey: they have a low tonal value. However, as we shall see later on, it is possible for a dark colour to have a high tonal value if it is illuminated by a strong light. Similarly, a light colour found in a shadow area may have a dark value.

In any painting or drawing, it is the contrast between light, medium and dark tones which creates the illusion of light falling on the subject. Even the most brightly coloured picture can look flat if it lacks tonal contrast.

△**Rhubarb** by Lucy Willis, watercolour, 15in×11in (37cm×27cm).
A full range of tones from light to dark can be seen here. The impact of tonal contrast can be best appreciated where the sunlit leaves stand out against the deep shadows, creating a space in which the rhubarb appears to come forward while the background recedes. When painting in watercolour, do not be afraid to use really dark tones when they are required. There can be a tendency to add too much water to the paint, producing pale, weak washes when dark ones could add strength and depth to the painting.

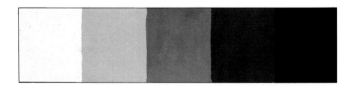

△The tonal value of a hue is determined by comparison with a scale of greys, ranging from white to black.

△**Marrow Seedlings** by Lucy Willis,
watercolour, 22in×30in (55cm×75cm).
'Tone' refers to how light or dark an
object or colour appears to be. Here, for
example, the flower pots have a much
darker tonal value than their reflections
on the table.

THE INFLUENCE OF LIGHT
ON TONE

When learning to judge tonal values many people have difficulty in allowing what they *know* about the tone of an object – that is to say its 'local tone' – to be contradicted by what they *see*. Knowing, for example, that a white cat is lighter than a grey cat can confuse and obstruct the artist when trying to assess the tonal values of the cats as they really appear.

The grey cat is darker than the white cat because grey absorbs more light than white does, but these so-called local tones of white and grey can change due to the influence of light and shade. In bright sunshine, for instance, the white cat will appear truly white only in those areas which receive direct light; those parts of its body which are in shadow will appear comparatively grey. Likewise, the grey cat might appear almost black where its body is in shadow, and almost white in those areas touched by very strong light.

TONE AND COLOUR

It is no different when looking at coloured objects. Each colour has its own tonal value, and the difference between a pale and a dark shade of the same colour – say, pink and red – is a difference in tone. A pink billiard ball and a red billiard ball are the same basic colour, but the pink one is light in tone and the red one is dark in tone.

However, when you shine a bright overhead light onto the billiard balls, their local tones are affected, and the top part of each ball – where it receives the light – acquires a higher tonal value than the underside, which is in shadow. The pink ball, which we know to be lighter than the red ball, could now appear the darker in certain areas: its shadow side may actually appear darker than the brightly illuminated top of the red ball. So you can see, very simply, how local tone is affected by light.

△ In these black-and-white studies you can see how the local tones of the grey and the white cats are altered by the effects of light and shade.

△**Kingcups in a Vase** by Pamela Kay, oil on canvas, 8in×10in (20cm×25cm). In this flower study it is the tonal contrasts which give each flower its delicate form. The light on the petals affects not only the tonal values but also the intensity of the colours.

HOW TO SEE TONE

Having understood the importance of tonal values and what they really are, the next step is to put this understanding into practice.

When you look at any object or scene you are confronted by a mass of visual information; your eye is bombarded by colours, shapes, tones and textures. In order to evaluate what you see, so that you can begin to make a picture without having a nervous breakdown, you have to train your eye to actually cut out a lot of the information it is presented with. In order to evaluate the tones in your subject, you must temporarily ignore texture and detail and think only in terms of light and dark.

A simple trick to help you to do this is to look at the subject through half-closed eyes. This will eliminate a large amount of distracting colour and give you simplified information about the tonal values of your subject. There is a real physiological reason for this, which has to do with the arrangement of the rods and cones in the retina of the eye, which perceive colour and light. Half-closing the eyes has the effect of blurring detail, which also enables you to evaluate tones more easily.

If you are undecided about a particular colour or tonal relationship, here's a useful method of isolating and comparing the colours: loosely grasp a pencil in your fist and then pull it out without allowing the resulting 'tunnel' to close up. This is your 'ever-ready viewer'. Hold up your viewer to within a few inches of one eye and close the other eye. You should see a tiny hole surrounded by darkness. Move your viewer around until you find the colour in doubt and you will almost immediately see it for what it is. If you then move the viewer to an adjacent colour, and then to the next, you can establish the first colour's true relationship to the others.

DRAWING IN PENCIL

When learning to see and evaluate tones, it is a good idea to start by drawing rather than painting. When drawing in pencil we are forced to think in terms of tone, translating light and dark colours into white, black, and various shades of grey in between. Often, in doing so, our monochrome drawings turn out to have a power and directness which a coloured painting sometimes lacks. (Perhaps this is why so many of our greatest photographers, too, have worked almost exclusively in black and white.)

In order to tackle a full-scale tonal drawing, remember the golden rule: start simple, then gradually work up to the complex. Begin by breaking down all the light and dark colours you can see into four main tones: white, light grey, dark grey and black. The white of the paper will provide the lightest tone in your drawing, while black is the darkest tone you can achieve with your pencil.

Having put down the four basic tones, you can now start to work into each area more of the subtle variations you can see.

▷**Wrapped Roses** by Lucy Willis, watercolour, 11in×15in (27cm×37cm). When I first considered this as the subject of a painting I saw a confusion of colours and reflections. By half-closing my eyes to simplify the mass of information into light and dark tones, I was able to make sense of what I was seeing.

◁**Charlie** by Lucy Willis, pencil, 22in×15in (55cm×37cm). Drawing in pencil is an excellent way to learn to use tonal values. Throughout an artist's career, tonal sketches can serve a number of purposes: as a means of seeing things in terms of light, as studies for subsequent paintings, or drawings simply to be enjoyed for their own sake.

luy willis 1987

Work over the whole drawing at once, rather than piecemeal fashion, and constantly relate one tone to another.

TONAL CONTRASTS

Remember to be quite clear about the tonal contrasts within your drawing; it's easy to over-emphasize the darkness of the nearly white tones in order to distinguish them from white. If these tones are too dark, and if you also exaggerate the lighter tones within the dark areas, the overall effect will be too 'jumpy'. The picture will lack punch, vibrancy and light. So if the tones are close in value, keep them close and thus retain the unity of the picture.

Whether you are drawing a simple apple or a far more complex composition like the one opposite, the same principles apply. Evaluating and drawing these fine gradations in tone can prove difficult, and if the drawing starts to go wrong – to look flat or disjointed – it is always best to go back to a simple four-tone approach.

One's choice of subject is often determined by an excitement about light and the way in which it falls. If this excitement seems to be fading away in the wrestle with tonal values, stop drawing, look again and try to regain that initial inspiration.

Project: tonal drawing

For this project, take a simple object such as a jug, and place it against an uncluttered background under a strong directional light. Make a series of tonal studies of the jug in pencil.

1 Using a soft pencil, make a light but reasonably accurate line drawing of the jug on white paper. Now pause to have a long, hard look at your subject through half-closed eyes before beginning to block in the tones of light and dark. Try to break down the lights and darks into four main tones: white, light grey, dark grey and black. Begin by plotting the darkest darks with heavy hatched strokes. Next, locate the lightest areas and, if necessary, draw a delicate line around the light shapes to isolate them. These lines will eventually be absorbed into the drawing.

1

2 Shade the entire drawing with a light grey, leaving only those white areas which you have isolated. This can be done quickly and loosely using a vigorous diagonal shading, which helps to avoid getting involved in detail.

2

3 All that remains now is to shade those areas which are tonally between light grey and black. Even with such extreme simplification, a satisfying sense of light and three-dimensionality can result.

4 Having completed your tonal studies in pencil, try tackling the same exercise in watercolour or ink. Notice in this example how the shoulders of the jug are defined by the darker background. Exercises like this are a great way of learning to see and apply tone.

3

4

TONAL KEY

onal key is a term used generally to describe the range of tones within a picture. It tells us about its overall lightness or darkness. It is useful to think in terms of tonal key when first starting to plan your painting, because this will help to determine its mood and emotional impact.

Certain objects are, of themselves, intrinsically 'high' or 'low' in key. A misty, sunlit morning with its pale brightness, or a woman in a white dress on a sandy beach, would be suitable 'high key' subjects; a darkened room at evening, or a dimly lit city street, would be 'low key' subjects.

A high-key painting is one which features mainly light tones and colours, and so gives a feeling of lightness and brightness. Generally, strong contrasts of tone are kept to a minimum, though some darker tones can be introduced to give the painting 'snap'. You will notice that high-key paintings are often suffused with light and convey a cheerful, bright mood or an atmosphere of softness and delicacy.

To capture a high-key mood in your paintings, use only the lightest tones in your palette of colours – and avoid black altogether. Remember that, even in the darkest of shadows, the tones will be lightened by reflected light on a sunny day; paint them too dark, and you will destroy the tonal unity of the picture.

Watercolours can be applied in a series of light, transparent washes to produce high-key paintings of great delicacy and subtlety. In oil or acrylic, the paint can be applied thickly, and the colours mixed with a high proportion of white if necessary. Be careful, though – overuse of white can produce a milkiness in the final painting.

A low-key painting is made up of tones from the darker end of the scale and can generate all those feelings connected with darkness. Some of the most beautiful paintings have been done in low key: they can be mysterious, sombre and powerful.

In a low-key painting there will be few brightly lit areas within the composition and what tonal variations there are will lie within the lower end of the scale. For the beginner, watercolours are less suited to this kind of painting because the mixing and application of strong dark tones can, if not properly controlled, lead to an unwanted murkiness. Oil and acrylic are ideal as you can manipulate the medium – overpainting, scraping away, blending and progressively darkening the tones – without danger of spoiling the painting.

◁**Dusk** by Arthur Maderson, pastel, 36in×28in (90cm×70cm).
Here the artist has observed the darkening evening landscape and produced a low-key painting using closely related shades of cool, sombre colour.

△**The Sunset Hour, Weymouth** by Arthur Maderson, oil on canvas, 57in×47in (145cm×119cm).
By confining his range of tones to the lighter end of the scale, Maderson has conveyed the impression of sparkling light. The painting is suffused with that all-pervading brilliance which is characteristic of the successful use of high-key colours.

△ **Twilight, Cookham** by Carlton
Grant, watercolour, 15¾in×23¾in
(40cm×60cm).

This atmospheric twilight scene is based
on the inter-relation of light and dark
tones.

TRANSPOSED KEY

The concept of tonal key becomes even more interesting when we want to creatively interpret our subject in order to get a particular effect, rather than painting it exactly as we see it.

Say, for example, you have before you a scene which is not of itself strongly 'high' or 'low' in key, but contains a range of tones from the top to the bottom of the scale of light and dark. In such a case it is perfectly possible to transpose all the tones either up or down the scale, in order to create a particular mood or lighting effect in your painting. For example, by transposing a middle-key scene of a winter landscape to low key, the cold, sombre mood of winter will come across more forcefully. Similarly, by transposing a middle-key river scene to high key, the effect of sparkling sunlight on the water might be shown to better advantage.

To make a musical analogy: if you consider the tones on the tonal scale as musical notes, then transposing the key in painting is rather like playing a tune an octave higher or lower than it was originally written. The relationship between the notes remains the same, but the sound communicates a different feeling and demands a different response.

In painting, then, transposing the tonal key leaves the tonal relationships intact, but alters the visual effects, and therefore the atmosphere of the painting. This can add enormously to the emotional impact of the picture.

◁**September Evening** by Arthur Maderson, oil on canvas, 48in×46in (122cm×117cm).
Although this interior is quite shadowy, Maderson has keyed his tones towards the higher end of the scale rather than the lower end. This, plus the use of juxtaposed greens, violets and blues, adds to the shimmering effect of light.

Project: pencil portraits

Find a friend who is willing to model for you, and make two quick portrait sketches in pencil. In the first sketch, transpose all the tones to a higher key; in the second, transpose them to a lower key. It is interesting to examine the different moods conveyed in these studies.

1 High key Using a hard (H) pencil, lightly block in the shadow areas with short, feathery strokes. Go no further than a middle grey – not black – in the darkest areas, and a light grey in the mid-tone areas. Leave white paper to represent the lightest tones.

In the finished drawing, notice how closely related the tones are; these subtle contrasts convey a sense of delicacy and a soft, pervasive light.

2 Low key Now draw your sitter under exactly the same lighting conditions as before. Just as you raised the value of the darkest tones in the first drawing, now try lowering the value of the lightest tones. Using a soft (6B) pencil, use hatched strokes to shade lightly over most of the drawing in order to establish the lightest tone. Build up with more hatched strokes to establish the mid-tones. Try to avoid using lines; allow the shapes and masses to emerge simply through the juxtaposition of tones.

As in the first drawing, the tonal range is limited, yet the impression of directional light is still apparent and the features are clearly defined.

PAINTING IN MONOCHROME

efore you can properly use paint to describe the effects of light it is necessary to understand how to paint tonally. People often assume that painting is all about colour, but many great artists have argued that a proper understanding of tone is equally essential.

To ease the transition from tonal drawing to painting in colour, it is a good discipline to attempt a painting in monochrome; it will help you develop a greater sensitivity to the role of light and dark in your paintings – in modelling form, creating mood, and of course, conveying the effects of light. By restricting your palette to various tones of one colour, you can go through the process of applying and manipulating paint in any chosen medium without having to worry about the complex considerations of colour. Your full attention can be given to capturing the effects of light through a tonal analysis.

The idea of conveying the illusion of light, space and atmosphere, without the aid of colour, is certainly a challenging one. You may find it helpful to consider this exercise as 'drawing with a brush'. Use the white of the paper or canvas for the lights, and the darkest tone of one colour for the darkest shadow areas. Used side-by-side, these two will provide the most dramatic contrasts. Try to restrict the tones in between these two extremes to no more than three to begin with; this will mean simplifying the many tones in your subject and grouping them into strong, cohesive areas of light and dark. It is this which will give your monochrome study a feeling of colour as well as light.

LOST AND FOUND EDGES

Resist the temptation to give equal emphasis to all the edges of the objects in your painting. You will lose the best of your picture if you do so, for this tends to create a series of 'trapped' shapes, like paper cut-outs, destroying the illusion of three-dimensionality. Instead, aim to create a subtle interplay of 'lost' and 'found' edges. A lost edge is one where the tones and colours between, say, an object and its background, merge together; the precise definition of the edge of the object becomes 'lost', or subtly implied. A found edge is strongly defined by contrasts in colour and tone between the object and its surroundings.

Look closely at the painting of the heron opposite. It is not entirely monochrome, but the subject itself demanded a limited palette of subtle greys. You can see the handling of the lost and found edges quite clearly. Notice, for example, the strong tonal contrast where the light strikes the front of the bird's throat. This is a 'found' edge. On the back of the neck, the tone is virtually the same as that of the background, only of a slightly different colour. This is a 'lost' edge. On the heron's back there are quite distinct areas of light and dark grey which contrast clearly with the tones behind. These, again, are 'found' edges.

It is this rhythmic interplay between lost and found edges which helps to describe the subtle effects of light upon a subject. It also helps to define three-dimensional form, and describes the advancing and receding planes of an object.

◁**Apples** by Lucy Willis, watercolour, 4in×7in (10cm×17cm).
In spite of the lack of colour, the apples are clearly defined in terms of tone. Monochrome studies like this are a useful means of learning to capture the effects of light without the distractions of colour.

▷**Heron** by Lucy Willis, watercolour, 30in×22in (75cm×55cm).
The form of the heron is described here through the counterchange of tones: light against dark and dark against light. In places where the tones are equal, there is often a subtle change of colour which helps to define the shapes and adds interest to the passages of 'lost and found'.

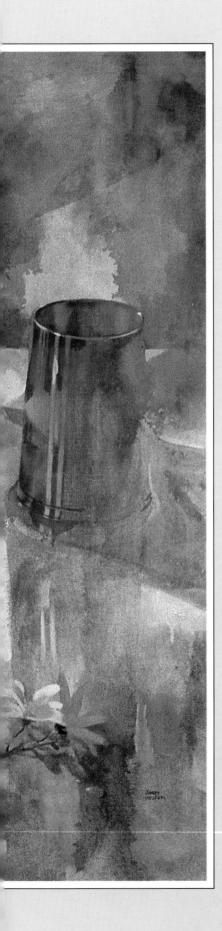

CHAPTER THREE

LIGHT AND COLOUR

L earning to see colour is one of the most exciting aspects of learning to paint. We know, on one level, what colour things are; but discovering the numerous ways in which these colours are transformed by light comes as a revelation. Light can modify all the characteristics of colour: its hue, its intensity, temperature and tone. The magical properties of light can induce colours to borrow from, and lend to, one another, producing subtle reflections and vibrant rhythms.

In the early 19th century, painters began to explore the way in which light and colour interact and the way that light can be expressed *through* colour in a painting. The Impressionists found that by placing small strokes of certain colours side by side, rather than by mixing them together on the palette, a remarkable vibrancy could be achieved. They used this technique of 'optical mixture' to spectacular effect, creating beautifully light-filled paintings.

In this chapter you will find guidelines for learning to see colour, demonstrations to show different ways of mixing and applying colours, and an exploration of the way in which complementary colours can be used to create and intensify the impression of light in your paintings.

Complementary Colours
And Light

When at school, most children become familiar with the three primary colours: red, yellow and blue. They learn that by mixing any two of these colours they can obtain the three secondary colours: orange, green and violet. What is often not taught is that each secondary colour is the 'complementary' of one of the primary colours. It is, in fact, the complementary of the primary colour that it does not contain. So green, consisting of blue and yellow, is the complementary of red; violet, consisting of blue and red, is the complementary of yellow; orange, consisting of yellow and red, is the complementary of blue.

It is important for a painter to be aware of this because of the extraordinary effect complementary colours can have on each other: there exists a particular dynamic tension between them which causes them to vibrate when they are juxtaposed. During the early 19th century, the Impressionist painters discovered that by juxtaposing complementary colours in this way they could achieve a remarkable, shimmering luminosity when describing the effects of light and shade.

The Impressionists rarely mixed their colours on the palette; instead, they applied the paint in small, broken strokes or dabs of pure complementary colours which, when seen from a distance, fused in the eye of the viewer. In this way an area of neutral grey, for example, could be beautifully rendered by working over the surface with strokes of blue and orange. These little patches of colour would appear quite separate on close examination, but when looked at from a normal viewing distance they would merge to create a grey which was both rich and luminous.

It is a good idea to experiment with broken colour and complementaries by making a selection of small colour samples. Work the whole surface of your sample with small strokes of two complementary colours, perhaps increasing in places the number of strokes of one colour in proportion to the other to see what effect it has. This can be done not only with pure primaries and their complementaries but also with, say, warm and cool greys or warm and cool greens. When you have become familiar with breaking up colour in this way, try applying this new approach to a simple composition. Remember to think of colour in terms of its temperature and

◁ **The Beach** by Andrew Macara, oil on canvas, 10in×12in (25cm×30cm). In this busy beach scene the artist has balanced passages of soft, neutral colour with carefully distributed spots of vivid colour. The neutral areas act as a foil for the bright colours, heightening their intensity by contrast.

▷ **Heading Home** by Sally Strand, pastel, 56in×36in (142cm×91cm). Near-complementaries – blues and yellows – are woven throughout this composition, creating a colour scheme that is lively yet harmonious. Note how the most intense complementaries are placed near the centre of the painting, providing a focal point around which the rest of the composition revolves.

to increase the proportion of warm or cool spots of colour where appropriate. When painting an orange, for instance, you may notice that the side in shadow becomes cooler as well as darker. Instead of mixing this cool dark colour on your palette, try juxtaposing small strokes of violet and blue, balanced with reds and oranges, until you achieve the result you require. The relationship of dark to light will remain the same, but the surface of your painting will become more lively and the effect of the colours on each other more dynamic.

Although you may not want to apply this technique as rigorously as the Impressionists or Pointillists did, experiments like this can inform all your painting and enhance your understanding of colour and light.

Study Arthur Maderson's painting *The Sunshine Stakes, Weymouth* (overleaf). Here you can see how the artist has used a complementary colour scheme – primarily cool blues and warm golds – and small, broken strokes to create a highly

atmospheric impression of the shimmering light of a late afternoon at summer's end.

Arthur Maderson usually times his trips to Weymouth to hit his favourite time of day, the sunset hour. The original idea for this picture dates back to a walk along the beach on a late afternoon in September, following a glorious, sunny day. It was packing-up time for those holiday-makers determined to squeeze the last moments of pleasure from what might prove to be the last moments of summer. It was a curiously poignant moment, with the day and the year sliding into decline: for a fraction of a second, Maderson sensed the sad aspect of late autumn and cold, dark afternoons to come. However, his spirits were suddenly lifted by the dramatic and beautiful impact of the final, glancing rays of orange sunlight which streaked low across the beach from behind the buildings along the promenade. As the quality of the light was changing swiftly, Maderson took a series of photographs and

◁**The Sunshine Stakes, Weymouth**
by Arthur Maderson, oil on canvas,
48in×48in (122cm×122cm).
In this painting, Maderson has created a
web of shimmering colour by juxtaposing
complementaries, which have the effect
of intensifying each other.

also made a rapid sketch, on which he made copious colour notes. He used descriptive colour images such as 'elderberry stain', 'burnt copper' and 'ripe damson', which would trigger his memory when, back at the studio, he began painting. His initial colour sensation was one of predominantly cool areas: of green/violet/blue shadows punctuated by the warm pinks and ochres of reflected light. Contrasted with this were the warm orange/ochre/yellow sunlit areas, laced with blue/violet in the shade.

The problem of responding to light in nature using pigments is a complex one. However, after years of what had seemed fruitless activity, essentially matching tonal values, Maderson saw a possible solution in a different approach: to attempt to graft together a purely intuitive response to colour in nature with a knowledge of colour interaction in a more theoretical sense. The artist explains: "I suppose that to a large extent I have accepted many of the ideas crucial to Pointillism, developed by Seurat and Signac. That is, I like to sub-divide colours into their constituent elements and apply them to the canvas in small strokes and dabs of pure colour which fuse in the viewer's eye. I try to energize the surface of the canvas by using colour to create muffled harmonies on the one hand, and complementary explosions on the other. This has the effect of producing, not more intense or vibrant colour, but colour which has appears more luminous. In nature one does not see a dull or non-luminous area: the relative shadow passages offer up a rich and subtle world of close colour harmonies which can be as delightful and rewarding as any burst of sunlight. A neutral grey can be charged into becoming a throbbing violet, or possibly a warm orange, by its juxtaposition with other colours."

The simplicity of the cool/warm opposition in *Sunshine Stakes* enabled Maderson to develop secondary harmonies of light and dark tone. For example, whilst being predominantly cool and shadowy, the foreground is relieved with light, warm orange/ochre notes. The central band of sunlit sand is reversed, being essentially warm and bright, yet broken up with cool blue-violet accents. The distance is a relatively complex area in which the various elements come together in a sort of visual square dance. As the artist explains: "Although I am now primarily interested in responding to light in terms of the organization of harmonious colour, my tonal training – how light or how dark one area is in relation to another – is still intact. It is this aspect which enables me to group my major family of shapes. If this skeletal element becomes too complex or fragmented it can destabilize, if not destroy, the overall structure of the picture. I feel more comfortable when I can break down my major areas to perhaps no more than four basic shapes."

Project: painting light through colour

Find a reproduction of a painting by Rembrandt, a painter who worked more than two hundred years before the Impressionists began their analysis of the effects of light and colour. Make your own interpretation of the picture using broken patches of intense, pure colour. See if you can achieve a vibrant, high key painting through the use of tone and colour, rather than using chiaroscuro alone, as Rembrandt did.

Look at Arthur Maderson's painting *Fragmented Sunlight*, in which he has interpreted Rembrandt's *Woman Bathing* in this way. You will learn a great deal from both painters.

▽ **Fragmented Sunlight** by Arthur Maderson, oil on canvas, 44in×38in (112cm×97cm).
The impression of light in this painting is created by the rhythms and patterns of broken colour, rather than by contrasts of light and shade. The whole composition revolves around the key relationship of yellow and violet, enriched with splashes of pink, orange and green.

△ **Woman Bathing in a Stream** by Rembrandt van Rijn (1606-69). Oil on canvas, 25in×19in (64cm×48cm).
As in many of Rembrandt's paintings, we find here an extraordinarily skilful use of chiaroscuro, in which the light-struck figure contrasts with the rich, dark background.

Colour Temperature

olour temperature describes how warm or cool a colour is. Red, yellow and orange are generally regarded as warm colours, while blue, green and violet are regarded as cool colours. However, each colour has its own degree of warmth or coolness. When compared to cerulean blue, for instance, cobalt blue is warm; alizarin crimson is a cool red, while cadmium red is warm.

However, what really determines the warmth or coolness of a given colour is the colour of the prevailing light, and of reflected light from nearby surfaces and the sky itself. Early morning light, for example, is quite cool. Late afternoon

sunlight, on the other hand, tinges every colour with its warmth (except in the shadow areas, which are correspondingly cool). Artificial light, too, has a colour identity; compare the warm, yellow light shed by a table lamp with the cool, greenish colour of fluorescent light.

So, instead of seeing things in terms of their local colour – a 'red' dress, a 'green' tree – learn to modify these local colours with the colour of the light. Observing the colour temperature changes caused by light is important in conveying a true impression of a particular place and time of day – as well as adding sparkle and vitality to your paintings.

◁**Cricket Match** by Andrew Macara, oil on canvas, 12in×16in (30cm×40cm). Although the local colour of the grass is the same green all over, its colour and tone are radically altered by the effects of sunlight and shadow. The artist uses cool, dark blue-greens and warm, bright yellow-greens to create a strong contrast between cool shadow and hot sunshine. Note how the grey shirts of the figures in the shade break into the sunlit area, thereby locking the composition together.

△**Villa Magia at Night** by Lucy Willis, watercolour, 22in×26in (55cm×65cm). It is not only the bright, warm colours in a picture which create the impression of light. Here the areas of cool, dark blue surrounding the house lend emphasis to the warm light spilling through the windows.

▷**Sunflowers** by Charles Sovek, oil on canvas, 24in×28in (60cm×70cm). Here is a painting which gives a strong impression of colour and sunlight. Yet large areas of the picture are painted with dark tones of neutral greens and blues; it is these which make the isolated splashes of pure colour so intense. Notice, in particular, how the brightness of the sunflowers is heightened by the subdued colours behind them.

▷**Warm Breeze** by Sally Strand, pastel on hardboard, 16in×28in (40cm×70cm).
Here is an example of how warm/cool colour vibrations can really make a picture sing. Strand began this painting by laying down washes that were opposite in temperature to those of her final pastel colours. For example, warm siennas and ochres vibrate as they show through and mix with the cool areas of blue pastel. Working with the end of the pastel stick rather than the flat side, she uses linear strokes that allow the colours to mix optically.

◁**Men in White** by Sally Strand, pastel, 23½×38in (58in×95cm).
The colour of shadows on white materials can be intensely beautiful. Sally Strand has observed and recorded a range of vibrant colours reflected and absorbed from the surroundings, which create a glowing luminosity. Even though the subject contains a lot of white areas, the artist has not used white pastel. Instead, she contrasts the light tones of blue, green, violet and yellow against the darker areas, so that they appear white.

Demonstration: White Still Life

The best way to observe the warm and cool colours that are present in 'white' objects is to set up an all-white still life and make a study of it. I set up this group in front of a north-facing window, because here the light would remain constant while I completed my study. I decided to work with pastel on a tinted paper, allowing the colour of the paper to act as a unifying element.

1 Using a white pastel stick on tinted paper, I sketch in the main outlines of the composition. I loosely block in the white of the table cloth and the dark area under the window frame with purple-grey and a touch of yellow ochre. This establishes the lightest and darkest areas of the painting. I then put touches of yellow, mauve and light grey on the shoes, bottle and window frame, pulling the shapes together and tightening up the drawing as I go along.

2 I darken the tones outside the window with a variety of warm and cool greys. I work continually from one part of the picture to the next, looking for correspondence and contrast in colour and tone. Light strikes the insides of the shoes, which I colour with warm yellows and browns, and I use mauve and blue for the outsides.

△**White Still life** by Lucy Willis, pastel, 12in×18in (30cm×45cm).

3 In the final stage I indicate the folds in the cloth with the same variety of warm and cool greys. I deepen the shadows and strengthen the lights with white and touches of pale yellow. The image is pulled into focus by sharpening an edge, accentuating a highlight, or merging colours where a softer effect is needed.

TONED SURFACES

Whether you are painting a landscape, an interior, or a still life, light is essentially the main subject of your painting. Remember: local colours are influenced by the colour of the prevailing light. One way to develop a harmony of light and mood is by working on a toned surface and allowing the colour of the surface to show through the finished painting, thus harmoniously unifying its overall colour scheme.

Choose a ground colour which is analogous to the lighting effect and the mood you wish to convey in your painting. For example, a golden-orange ground colour will set the tone for a painting of warm afternoon light, while a muted blue or grey will help to establish the mood of a cold, wet day. The lightness or darkness of the ground colour will also influence the picture's mood, so make sure you pitch it right. The safest bet is to choose a mid-tone between the lightest and darkest colours in the painting; that way, you can work up to the lights and down to the darks.

Painters in oils and acrylics may already be familiar with the idea of tinting the canvas with a thin wash of colour prior to painting, and pastellists have at their disposal a vast range of tinted papers. But even transparent and semi-transparent media such as watercolour and gouache can benefit by being applied to a tinted paper; the colour of the paper shows through each stroke and imparts a quiet harmony that is otherwise difficult to achieve. It is well known, for instance, that some of the great masters of watercolour painting used cold tea to tone their papers a pale golden brown.

Demonstration: Picking Flowers in the Douro Valley

For this colourful scene I worked on a neutral buff paper, so that each pastel colour that is applied is shown to its full potential. Starting with the focus of attention, the figures, I have from the beginning related them to the surrounding landscape colours, which either echo or complement them.

2

1

48

3

△ Picking Flowers in the Douro Valley
by Lucy Willis, pastel, 15in×22in
(37cm×55cm).

1 I draw in the shape of the hat with a
pale yellow pastel, then position the
figures by sketching in their clothing and
hair. I try to see the figures as simple
shapes – of pink, mauve, black, or brick
red – rather than as people; this helps me
to assess proportion, colour and tone
instead of getting bogged down in detail
at this early stage. I then begin to sort
out the vast expanse of landscape behind
by plotting a few landmarks in dark
green, white, and a mixture of blues
blended with white.

2 As I gradually define the trees and
bushes with brighter greens, I also return
at intervals to add light and form to the
figures. This is done with white for the
highlights on hair, hat and arms, and by
adding strokes of different reds and
purples to darken and solidify the forms.
Some of these colours are at the same
time introduced into the landscape as
dots and dashes to represent flowers
and trees.

3 The distance dissolves into a haze of
light, which I represent by blending
quite a lot of white into the paper with
my fingers. With a few more strokes of
muted green I suggest the trees and
reflections on the far side of the river.
While sorting out the confusion of detail
in the middle distance I occasionally find
I am getting too fiddly and have to blur
out the offending passage and re-state it
more clearly and freshly. Rather than
cover every inch of the paper I leave
quite a lot of its warm colour in the
foreground to accentuate the greens in
the distance.

CHAPTER FOUR

SHADOWS AND REFLECTED LIGHT

It is not only the pale, bright areas in a painting which create the effect of light, but also the dark, contrasting passages of shadow. In fact, it is often true that without the shadow areas, the sunlit areas definitely lack sparkle.

Shadows are endlessly varied in shape, colour and tone, and can become as much the subject for a painting as the solid world upon which they fall. This chapter looks at ways of painting shadows, and at their various properties: the quality of their edges, the intensity of their colour and tone, how they absorb and reflect colour and light, and how they help the artist to describe surfaces and model form.

HARD AND SOFT SHADOWS

It's worth taking the time to make a study of shadows; there's more to them than meets the eye. For instance, why do some shadows have hard edges and others soft? One reason has to do with the actual source of light. If there is a single, bright light source, such as the sun or a desk lamp, the shadows cast by the object upon which the light falls will be strong and clear. If, however, the light comes from a large, diffuse source such as a window away from the path of direct sunlight, or the sky on an overcast day, the resulting shadows will be soft and blurred at the edges.

Another factor has to do with the distance of the object casting the shadow from the surface upon which the shadow falls. It may help you to think of the shadow cast by a flagpole on a sunny day. You will see that where the shadow lies nearest to the base of the pole the edge will be sharpest. The further the shadow travels from the flag-pole, the softer it becomes, because more light is reflected into it from the sky.

Next time there is a sunny day, try looking for hard and soft shadows outdoors and resolve to recreate them in your paintings. They will not only add poetry to your paintings, but also give a more convincing illusion of light.

▽ **The Piano Duet** by Lucy Willis, watercolour, 15in×22in (37cm×55cm). Here the shadows are not cast by direct sunlight but by diffuse light from a large window in the background, so they are soft and slightly blurred. Note how the shadows vary in tone depending on how near they are to the objects casting them. The legs of the stool, for example, cast shadows which gradually fade towards the foreground.

▽**Sunlight and Shadows** by Andrew Macara, oil on canvas, 12in×14in (30cm×35cm).
The poetry of this painting lies in the treatment of the shadows cast on the wall. Those of the fence and bush are clear and sharp, while that of the more distant tree is paler and softer. Note how the blurred shadow of the tree's canopy indicates movement.

◁This detail from the top centre of the painting shows how the blurred shadows of the tree's foliage are achieved by softly blending the shadow colour into the surrounding area. The shadows of the bush are more sharp-edged.

Project: shadows and mood

The aim of this project is to demonstrate the effect that the prevailing light has on shadows, and how this in turn can affect the mood of your painting.

Set up a very simple still life directly in front of a window which receives a fair amount of light – preferably in a place where you can leave the group undisturbed for a few days if necessary, because you are going to paint two versions of it.

Paint the first version of your still life on a bright, sunny day (choose any medium you like). Paint the second version in the late afternoon or early evening, when the light is softer and more diffuse. When both paintings are completed, compare the results and notice how the atmosphere and the composition have been affected by the different kinds of lighting.

1 Strong light Bright, crisp sunlight shining directly through the window casts sharply defined areas of light and shadow which are as important a part of the composition as the objects themselves. As we saw earlier in this chapter, shadows cast by nearby objects tend to have sharper edges than those cast by objects further away: note here how the shadows cast by the objects on the table have sharper edges than those cast by the window.

1

△**Morning Break I** by Lucy Willis, oil on canvas, 11in×10in (27cm×25cm). Strong, clear shadows cast by direct sunlight become important factors in the composition.

▷**Morning Break II** by Lucy Willis, oil on canvas, 11in×10in (27cm×25cm). In contrast, the shadows here are soft and unobtrusive, gently modelling the forms and absorbing reflected light.

2

2 Diffuse light In the second painting the mood is much quieter and more restrained. The same objects are lit this time by a warm, soft evening light which bounces off the colour of the back wall and creates two sets of subtle shadows: those blue-grey ones which come towards us and orange-tinted ones that fall to the left. Because the light is diffuse the tonal contrasts throughout the composition are much less marked, and all the shadows are soft-edged. To achieve this effect, blend the wet oil paint into the surrounding areas with your finger or a dryish hog's hair brush.

THE COLOUR OF SHADOWS

Early painters used to create shadow simply by adding black pigment to darken the paint. Then, during the 19th century, physicists began to make exciting discoveries about the constitution of colours and the structure of light. Eugène Chevreul, the French chemist, observed that shadows are often tinged with colours complementary (opposite) to the colour of the object casting the shadow. Chevreul's theories had a great influence on the work of the Impressionist painters, and these painters began to look into shadows and see there a great wealth of colour. Because of the way the Impressionists have taught us to see, it is no longer unusual to discuss shadows in terms of colour.

It is, in fact, quite possible for the colour of shadow to be simply a darker version of the local colour of the object in shadow. But where there is bright, all-pervasive light, as you would find outdoors on a sunny day, shadows tend to pick up the colours around them. Most shadows contain some blue, for example, which is reflected from the sky. In addition, as Chevreul discovered, shadows often contain some of the complement of the colour of the object casting the shadow. Hence, the shadow on a red apple's surface will be tinged with green. So, when painting the cast shadow of an apple, you might select a dark red, a blue and a green (each of the same tone) and layer or blend them to create the colour you want. As an artist, you'll find that painting colourful shadows will make your pictures come alive.

△ **Point of Sunset, Dordogne** by
Arthur Maderson, oil on canvas,
32in×44in (80cm×110cm).
Arthur Maderson has captured that
moment, at the end of a summer's day,
when the cool colours of the enclosing
dusk are just beginning to take over from
the last rays of warm sunlight. The
foreground shadows consist of patches of
greens, blues and violets which the artist
has broken up with intense, warm
colours to give the impression of sunlight
glinting through the grass and trees.

◁ **Table in the Garden with Geraniums**
by Pamela Kay, oil on canvas,
10in×14in (25cm×35cm).
The impression of light in this painting is
created almost entirely by the shadows.
Notice how the colourful shadows on the
white cloth make it a dynamic element in
the composition.

△ **Baby Cup and Saucer on White
Embroidered Cloth** by Jacqueline
Rizvi, watercolour and body colour,
9½in×12½in (23cm×31cm).
A soft, diffuse light from the left creates
delicate shadows which tell us everything
about the subject, from the shape and
form of the cup and saucer to the folds
and embroidery of the cloth. Note the
subtle use of both warm and cool greys in
the shadows.

Painting Luminous Shadows

Having discovered that shadows can actually be colourful, let us now turn our attention to the *way* in which we paint them. Very often, beginners fall into the trap of over-mixing their shadow colours, creating a dull, uniform grey that somehow looks flat and dead. As we have already seen, most shadows contain some variation of tone and colour, due to the presence of reflected light; don't be afraid to apply your colours loosely on the paper or canvas, in order to recreate this effect.

When painting shadows in oils or acrylics you can manipulate the paint to give a wide range of effects. You can apply loose strokes of warm and cool colour to create luminosity; you can scumble, scrape, wipe and blend with a knife, brush, rag or finger; or you can apply transparent washes of dilute paint one on top of the other. In pastels, too, you can apply different coloured strokes side-by-side, then layer and blend them to weave a mesh of rich, vibrant colour.

In watercolour, beautiful effects are achieved by allowing colours to fuse wet-in-wet, and by lifting out patches of colour with a damp brush to indicate reflected light.

Glazing

Another way to achieve a great depth of luminosity in shadows is by glazing one layer of transparent colour over another. In oils this can be a lengthy process, as you have to wait for one layer to dry before you can apply the next. With acrylics, however, a similar result can be achieved much more quickly as the paint takes far less time to dry. It is interesting to experiment with glazing techniques in order to see just what effects you can create. Try painting the first layer of the shadow area with an unlikely colour, say vermilion. Overlay this with successive layers of cool greens and blues (diluted to a very thin consistency) until the appropriate neutral colour results. If your glazing has been

successful, the layers of colour will take on a deep 'glow'.

A similar technique can be used in watercolour: one wash can be applied over another (dry) underlayer, like sheets of coloured tissue paper, to build up a deep tone in the shadow. This must be done with great care; make sure that each wash is thoroughly dry before applying the next, and put the paint on cleanly and freshly. Try to avoid stirring your brush around on the paper and disturbing the colour underneath, as this will only result in muddy, dead-looking shadows and not the glowing richness you are aiming for.

DIRECT PAINTING

The obvious advantage of using the method of glazing or overlaying in watercolour is that you can build up slowly from light washes to the required depth of tone and colour. However, if you are bold and decisive, it is possible to lay down equally successful shadows in one fell swoop. Some artists prefer to work in this way because there is less risk of overworking, which is the bane of watercolour painting.

When mixing the colour for the shadow make sure your paint is dark enough for a single application; it must not be too

△ **Pinnacle Rocks Near Swanage** by Ronald Jesty, watercolour, 13in×13in (32cm×32cm).
In this unusual and striking composition the central reflection is a key element. The artist has skilfully overlayed several layers of transparent grey wash to give a variety of subtle colours and tones without sacrificing the clarity and freshness of the paint.

▷ **The Hat and the Fiddle** by Lucy Willis, watercolour, 15in×11in (27cm×37cm).
If you look at the shadow of the violin you will notice a slight warmth in the colour at the base, which is where I started my wash. As I progressed upwards I cleaned my brush and quickly added a bluer mixture, which blended with the first colour.

◁ **Gentleman Caller** by Joan Heston, oil on canvas, 32in×42in (80cm×105cm).
The artist was interested in the quality of light falling on colourful, reflective surfaces. The luminosity of the shadows is achieved by glazing thin, transparent layers of paint on top of each other. The colours used were sap green, ultramarine blue, bright red, cobalt blue, and yellow ochre, mixed with a glazing medium.

dilute (remember that watercolour dries much lighter than when first applied). If you are not quite sure, test the tone of your mixture on a piece of scrap paper before you start. Look for any variation in colour or tone within the shadow and prepare other colours if necessary. As you want your wash to be continuous and flowing there will be no time to do this once you have started.

First, analyze carefully the shape, colour and tone of the shadow in question. Decide whether it is to have a soft or a hard edge, as this affects the way the paint is applied from the beginning. If the shadow is clear and sharp, the entire shadow shape can be applied in one continuous wash with the edge left to dry crisply. If, on the other hand, the edge of the shadow is soft and fuzzy, the paint should be allowed to run into the surrounding wash. If the shadow happens to fall on a white surface, then the edge of the wet shadow area can be blurred by painting a rim of clean water around it into which the paint can run. You can also dampen the entire shadow area before you start to paint it and then charge your shadow colour into it 'wet-in-wet'. In both cases it is important to control the amount of water on the paper, otherwise you will find the paint collecting in pools as it dries.

◁ **Mrs Maudsley's Greenhouse** by
Ronald Jesty, watercolour, 15in × 11in
(37cm × 27cm).
In this wonderfully inviting picture it is
the contrast between the intense, dark
shadows and the bright sunshine which
gives the strong impression of light and
depth. The telling spots of dappled
sunlight which break into the shadow on
the wall heighten this effect and entice us
into the illuminated greenhouse.

△ **Sun on Clear Creek** by Doug
Dawson, pastel, 20in × 40in
(50cm × 100cm).
Here the entire foreground is thrown into
shadow, providing a powerful foil for the
morning sunlight in the distance. Even
though the foreground is in shadow,
however, there is still detail and
luminosity, due to the influence of
light reflected from the sky.

REFLECTED LIGHT

The study of shadows is not complete without a discussion of reflected light, for the ability to capture reflected light in your paintings is an important factor in creating atmosphere and luminosity.

Reflected light is most easily seen on a bright, sunny day, and is caused by coloured light bouncing into the shadows from the sky or from nearby objects. Shiny, light-coloured objects reflect the most light, but everything reflects light to some degree.

In *Rug and Hat* (below) reflected light is clearly seen on the hat as an area of pale yellow within the shadow side. This light has been reflected from the surrounding surfaces – the rug and parts of the brim of the hat – which are in direct sunlight. In addition, notice the faint glow of yellow just above the blue ribbon where it lies upon the rug. This is the reflected colour from the hat falling within the blue shadow of the rug.

When painting reflected light, resist the temptation to overemphasize its tone and colour. An area of reflected light can never be as high in tone as an area lit directly, but it will be lighter in tone than the shadow within which it falls. In *Rug and Hat*, for example, the reflected light is brighter than the shadow, yet darker than the bright rim of the hat. It is the very subtlety of these tonal changes that often gives a painting a particular, luminous quality.

▷**Magnolia** by Lucy Willis, watercolour, 15in×11in (37cm×27cm). Here the fragments of light reflected on to the wall of the house are indicated by lifting out the colour with a clean, damp brush. Much stronger tonal contrasts are used to paint the light reflections in the window.

◁**Rug and Hat** by Lucy Willis, watercolour, 11in×15in (27cm×37cm). Reflected light can be seen here as a yellow glow in the shadow on the left side of the hat. The effect was created by surrounding this patch of cadmium yellow, and a touch of red and violet, with an area of cobalt blue and mauve, allowing the two washes to run together.

△**Vanessa** by Lucy Willis, oil on canvas, 10in×12in (25cm×30cm). Notice how reflected light plays upon a complex form such as the human face. In this portrait the entire face is in shadow and only the hair and the tops of the arms are in direct sunlight. Within the shadowy face there are numerous colours and tones. Warm yellow light reflected from the clothing strikes the downward-facing planes: under the bottom lip, under the nose, under and between the brows. At the same time a cool, blue light from the sky is reflected by the upward-facing planes of the nose and the top of the forehead. Reflected light, by picking out certain planes of the face, helps to model its form and captures the luminosity of the skin.

How Light Models Form

The way light and shade fall on a particular object contributes largely to the description of its volume and solidity. In a painting, therefore, the illusion of three-dimensional form is achieved through careful observation of the contrast of light and shade. Imagine, for example, how light strikes the cubed shape of a box, and the spherical shape of a tennis ball. The change from light to dark on the box is clear-cut and so describes an abrupt change in plane. On the ball however, the gentle gradation from light to dark describes the smooth curve of the surface.

The more visual information you can give, the more effective will be the illusion of solidity. When you consider reflected light as well as direct light falling on an object – on the ball for example – the picture becomes more complete.

In the painting of the lorry (opposite) the curved surface of the billowing tarpaulin is described by a gradation of tone. At the nearest corner of the lorry, on the other hand, there is a sharp change in tone, just as with the cube, which describes the sharp change in plane. Finally, the sense of volume is accentuated not only by the way the light falls but also by the ropes and the white stripe which act as 'contour lines' which accentuate the overall shape.

◁ **Bread Rolls with Carnations in a Vase** by Pamela Kay, oil on board, 10in×10in (25cm×25cm).
By skilfully blending and scumbling the paint, the artist gradates the colours from light to dark and achieves a satisfying sense of roundness in the bread rolls, as well as a convincing hollowness in the cups.

R.Jesty '87

△ **Mussel Shells and Lemon** by Ronald Jesty, watercolour, 5in×7in (12cm×17cm).
The impression of solidity is so tangible in this painting that we can almost reach out and pick up these objects. Closely observed patterns of light and shade describe both the concavity of the upturned shell and the sculptural forms of the lemon and peel.

◁ **Truck** by Lucy Willis, oil on canvas, 40in×48in (100cm×120cm).
Notice the sharp contrast between light and shadow where one plane meets another on the sides of the truck; compare this with the gradual transition from light to shadow on the rounded form of the tarpaulin.

Demonstration: Sarah on the Sofa

When painting people it helps to have an understanding of the way in which light models form. By paying attention to the play of light and shadow on the various surfaces of a figure, you can achieve an impression of solidity and roundness. In this subject I had plenty of nice, rounded forms: arms, legs, cheeks, and a comfortable sofa. The light in the room was diffuse, so the shadows were soft and graduated. I worked in watercolour, for speed and convenience, as I knew that my young sitter would not allow herself to be captive for very long.

1

2

3

1 I start by mixing the basic flesh colour from cadmium red, cadmium yellow and a small amount of cobalt blue to take the edge off the resulting orange. I use this to draw in the position of the arms, legs and face. Even in this first series of shapes I try to begin to model the form by darkening the tones in some places – under the knees for example – and leaving highlights in others, as on the forehead and knees. I mix a bold, dark colour for the hair – Prussian blue added to Vandyke brown – and this completes the shape of the head and plots the position of the ear. It also establishes the darkest tone in the painting, to which all the other tones can be related.

2 Using a slightly darker shade of the same flesh mixture, I start to emphasize the roundness of the limbs and the face, adding a little more cadmium red to the cheeks. Starting with the colour now on my palette, I mix in alizarin crimson and use this for the cushion behind as well as for the mouth and nose. I find that the remains of one colour invariably get used in the beginning of the next, and this ongoing cycle can help to create a satisfying harmony of colour.

3 The blue-grey shadows on the dress go in next, adding solidity to the body. The various planes of the sofa are described with simple underwashes of shadow before the pattern is applied. I find pattern, as well as tone, can be enormously helpful when modelling form; simply flattening the shapes of the flowers on the seat gives the impression of a horizontal plane.

4

△**Sarah on the Sofa** by Lucy Willis,
watercolour, 15in×11in (38cm×28cm).

4 Taking a long, quiet look at the painting, I realize that the shape of the head is slightly wrong. A sliver of dark hair at the back of the head – so confidently put in at the very beginning – now has to be removed. This kind of adjustment can be a nightmare in watercolour. However, I take a brush full of clean water and dissolve the dark paint, dabbing and lifting with a tissue until I am satisfied. When the area is completely dry, I carefully repaint the pattern of the fabric in that area. A few more fine adjustments of tone on the head and the background, and the painting is finished.

67

Susie by the Front Door by John Ward, oil on canvas, 30in×20 (75cm×50cm)

CHAPTER FIVE

LIGHT OUTDOORS

It is an exciting but perhaps daunting moment when you first decide to take your easel outdoors and paint nature 'on the spot'. How do you begin? Whether you are sitting in your own garden or perched on a mountain top, you will have to cope with the vagaries of the weather, the constantly changing light – and very often a moving subject, too. However, outdoor subjects can be fascinating and challenging, and overcoming these obstacles can provide the impetus for really exciting work.

This chapter presents nature in a variety of moods, in all seasons and weather – rain, storm, mist and bright sunshine – and shows you how to capture these fascinating phenomena of light and atmosphere in your paintings.

CHOOSING A SUBJECT

It is very easy to get carried away with the idea of 'going out painting', and the exciting prospect that out there somewhere the perfect picture is just waiting to be painted. In my experience, though, the reality is very different. No sooner do you come across a potential subject than you begin to wonder whether there might not be something just a little more interesting, unusual or better lit, just around the corner. And so it goes on. Eventually you return home, tired and dissatisfied, with precious little achieved. It is rather like going Christmas shopping without having a clear idea about which presents you want to buy.

One way to avoid disappointment is to have a clear idea of what you wish to paint *before* you set out. Often you will notice good subjects for paintings while going about some other business; you can store up these ideas for future use, even though it may be a long time before you can take the opportunity to make them into a painting.

The French Impressionist painter Camille Pissarro wrote in 1893: "Blessed are they that see beautiful things in humble places where other people see nothing! Everything is beautiful, all that matters is to be able to interpret." In other words, it is not your subject but what you see in it that matters. When making your choice of subject, try to avoid being too ambitious at first. The grand view may be very tempting, but you are more likely to find success if you start on a smaller scale. There can be just as much of interest in your immediate surroundings as there is further afield. In a corner of a garden, a park, or a street you will find, once you begin to look, a great many appealing compositions. I find my own vegetable garden a constant source of inspiration, particularly on sunny days when the light transforms a humble cabbage patch into a symphony of intense greens, blues and yellows.

◁ **Beach Series IV** by Sally Strand, pastel, 36in×55in (90cm×138cm). When painting outdoors, anything can be your subject; all you need is the willingness to look at your surroundings with a fresh eye. In this pastel, Sally Strand has focused on 'an intimate glimpse of the business of living.' She has captured the elegance of gesture, composition, and light that can encompass even the most ordinary situation.

▷ **Offenham Earlies** by Lucy Willis, oil on canvas, 24in×20in (60cm×50cm). You do not have to go far from home to find inspiration for your paintings. I find my own vegetable garden a constant source of ideas.

COPING WITH THE LIGHT

It is light, of course, which makes even the most everyday subjects so rewarding to paint. Outdoor light, though, is neither entirely predictable nor controllable. Painting on an overcast day can prove easier than on a bright, sunny day because you do not have strong contrasting shadows to chase after and you may be able to work for far longer before the light changes.

On a sunny day you have to learn to work within a new set of considerations. The sun moves across the sky – at an alarming rate, it seems, when you are absorbed in your work – and shadows throw themselves unexpectedly across your subject. It is difficult enough to paint shadows when they are still; how do you cope when they move around?

There are a number of ways in which you can overcome this problem. One is to work fast so that your painting is completed by the time the light has started to change. When I

first started to paint landscapes and had difficulty with this particular problem, my father gave me some very small pieces of oil paper, a few tubes of paint, and a large paintbrush, and let me loose on the local landscape. It was a good prescription which cured me of the tendency to get bogged down in detail, forced me to see the landscape in terms of light, and enabled me to complete each painting quickly.

Although this was, perhaps, an extreme remedy, it worked, and I still bear in mind the idea of simplification and the necessity of working fast when painting outdoors. The same method – working quickly and on a small scale – can be used when making sketches or studies of your subject prior to painting. Indeed, making monochrome sketches is always a good idea, because they are a means of distilling the essence of the scene and 'editing out' any unwanted details which might clutter up your painting. You can take these sketches back to

the studio and work from them indoors, which is especially helpful if you are unable to return to the same spot at the same time another day.

Once you get absorbed in a painting outdoors, you may suddenly look up and realize that the shadows have changed and that a lovely area of reflected light, for instance, has gone. To avoid this frustrating experience, it is often worth painting the shadows first and building the rest of the picture around them. If it is possible to complete the painting in a couple of hours or so, you should be able to cope with the changing angle of the sun in this way. But if you are attempting a more ambitious composition which requires many hours of work, don't be tempted to continue painting for any longer than this. I have seen some very strange paintings, in which the light appeared to be coming from all directions at once! It is far better to do as Monet did, and return to the painting at the same time each day, when you know that the direction of the shadows will be the same.

◁ **Scything the Grass** by Lucy Willis, watercolour, 15in×22in (38cm×55cm). The sun was shining directly toward me when I started this picture. I worked quickly so as to keep up with the drying paint and the changing shadows. After about two hours I had to finish as the mist had lifted in the distance and the changing position of the sun had transformed the scene.

△ **The Three Graces** by Charles Sovek, oil on canvas, 22in×28in (55cm×70cm).
Note the broad handling of the paint here. Aside from the heads and hands there's very little detail, yet the impression of strong sunlight is very effective. There's a lot to be said for working quickly and spontaneously, to catch the 'first impression'.

THE PRACTICAL PROBLEMS

Painting outside in the sunshine can be an extremely enjoyable experience, but there are certain practical considerations to be taken into account. One of the drawbacks is that you can be dazzled by the very light which you have gone out to paint, making it difficult to judge tones and colours accurately. If you don't want to spend the whole time squinting, it is a good idea to wear a wide-brimmed hat to cut out the sun and most of the glare from the sky. Sunglasses are not usually to be recommended as they can radically subdue colours and tones. If at all possible, find a spot in the shade, though even here you may need to wear a sun hat if your evaluation of tones and colours is disrupted by the glare from the sky.

PAINTING IN WATERCOLOUR

Watercolour is a fine medium for painting out of doors because its fluid nature allows you to capture the fleeting effects of light and atmosphere spontaneously. In addition, only a small amount of equipment is needed: a paint box, brush, paper and water jar.

The essence of handling watercolour is learning to control the speed at which the paint dries. On a breezy, sunny day it can dry too fast, giving hard edges where you do not want them, making it difficult to achieve large areas of smooth, unbroken wash. It is essential, therefore, to work decisively, since there will be no time to stop and change your mind about a particular tone or colour once you have committed brush to paper. Once the wash has dried to an unwanted hard edge, no amount of remedial work will retrieve the situation, for the clarity and freshness of watercolour is lost when you have to re-work it.

If, for example, you have a large expanse of sky to paint in a single wash, spend plenty of time considering which colours to use. Then make sure you mix large enough quantities of these colours to enable you to lay down the whole sky without having to re-mix in the middle. You will obviously be better off using a large brush for this, but even on smaller areas a fairly large brush, providing it has a good point, is to be

recommended because it holds a lot more wash and enables you to work quickly.

In damp or humid conditions, the problem is that washes can dry too slowly, so that they run together in an irritating manner when you wanted to keep them separate. Here, patience is a virtue. You must allow one wash to dry before adding the next – unless of course you wish to experiment with the expressive qualities of wet-in-wet washes.

PAINTING IN ACRYLIC

Painting in acrylic has the advantage that you can overpaint as much as you like. While it can be diluted with water and used in a similar way to watercolour, it is more usually applied thickly, like oils. Acrylic paint dries very quickly, especially in a warm, dry atmosphere, which can be both a help and a hindrance; it enables you to work over a previous layer of paint almost immediately, but it sometimes makes the paint difficult to manipulate, both on the palette and on the canvas, because it solidifies so quickly. The main advantage of acrylics over oils, however, is that your picture will soon be dry enough to pack up and take home without risk of smudging.

◁ **The Market, Rialto, Venice** by Arthur Gerald Ackerman R.I. (1876-1960). Watercolor, 6⅞in×10in (17cm×25cm).
There is a powerful feeling of light and vitality in this free and easy watercolour. The warm, colourful shadows have been painted quickly and decisively, leaving small fragments of bare paper to describe the sunlight striking the figures. Notice how the artist has used the full range of tones, from white to almost black, to give the painting depth and strength.

△ **Sutherland Coast** by Aubrey Phillips, pastel, 20in×28in (50cm×70cm).
Pastel is a marvellous medium for capturing the fleeting effects of light outdoors, because it is so quick and responsive. Here the artist has allowed the colour of the paper to establish the overall tonality of the scene. Quick, calligraphic strokes and blendings give an impression of a blustery day.

PAINTING IN OILS

Painting with oils out-of-doors can be extremely rewarding. One of its principal joys is that it is a forgiving medium, staying wet long enough to allow you to manipulate it, make alterations, overlay it, or even wipe out whole passages. However, once again it helps to be decisive and not to rely too much on overpainting if you want to retain freshness of colour and liveliness of brushstroke.

Of course, with oils you will need much more in the way of equipment, and you will return home at the end of the day with sticky paintings which inevitably manage to fall face-down on the ground or on the backseat of your car. Still, I always think of Monet pushing his old pram loaded with easels, canvases, paints, and brushes, heading undaunted into the countryside. You only have to look at his paintings to realize that it was well worth the effort.

To make your outdoor painting quicker and easier, it is a good idea to tint your canvas with one overall colour and allow it to show through the finished painting. This has three distinct advantages: it cuts down the white glare of the canvas

and makes it easier to evaluate relative tones and colours; it helps to establish the overall mood and tonality of the scene; and the repeated traces of colour have a harmonizing effect on the colour scheme of the picture. The canvas can be prepared the day before by rubbing a quantity of colour, thinly diluted with turpentine, all over the primed canvas. I usually use pigment left over on the palette at the end of a day's painting, thoroughly mixed and diluted with a few brushfuls of turpentine. This yields enough nice, runny, neutral grey or brown to cover a small canvas for outdoor painting.

Some painters prefer to work on a less neutral, more vividly coloured ground which contrasts, rather than harmonizes, with the predominant colours in the landscape. This is best achieved by using complementary colours. For example, if you are about to paint a landscape in which cool greens and blues predominate, you might choose to tint your canvas with a warm reddish brown. This complementary contrast has the effect of enriching your painting by making the greens and blues appear more intense.

◁ **Wild Lavender, Dordogne** by Arthur Maderson, oil on canvas, 28in×20in (70cm×50cm).
Here the artist has used a thick impasto to build up a richly textured surface. The palette of vivid violets, blues and spots of orange, as well as more neutral greens, lends vibrancy to the meadow and trees.

△ **Antibes, Evening** by Nigel Casseldine, oil on gesso, 9in×10in (22cm×25cm)
In contrast to the painting opposite, here the paint has been used more thinly and smoothly. The ochrish underpainting shows through on the roof of the building and in the other dark passages, providing flecks of warm, vibrant colour that complement the blues and greys.

DIRECTIONAL LIGHT

There can be few things more enjoyable than going out to paint on a warm, sunny day. Most of us, however, look for a subject which inspires us without necessarily analyzing the particular way in which it is lit. The direction and intensity of the light affect not only the impression of form and substance, but also the atmosphere and mood of a scene. It can be interesting and instructive to study the same subject at different times of the day and compare the effects of light falling onto it from different directions.

△ **Roadside Café** by Moira Clinch, watercolour, 18in×12in (45cm×30cm).

THREE-QUARTER LIGHT

When light strikes the subject from an angle of 45 degrees or so, it creates a strong impression of form, volume and texture. The tonal range is wide, and the resulting image has plenty of vitality. In Moira Clinch's painting *Roadside Café* (above), the sharp contrast of light and shade between the front and the side of the building contributes to the strong sense of perspective. In addition, the deep shadows and bright highlights give an impression of the heat and glare of midday in New Mexico.

△ **Girl in a Straw Hat** by Lucy Willis, watercolour, 15in×22in (38cm×55cm).

SIDE LIGHT

Light from the side, like three-quarter lighting, gives a strong feeling of solidity and three-dimensionality. In *Girl in a Straw Hat* (above), the contrast between light and shadow, especially on the girl's torso, describes her form and places her at right-angles to the plane of the beach. The illusion of solidity created in this way brings the figure forward in the picture plane. The strong cast shadows produced by side light are also very descriptive, helping to define the contours of the surfaces upon which they fall. Notice here how the shadow cast on the towel anchors the figure to the ground upon which she is sitting.

△ **Study of Sunset** by Doug Dawson, pastel, 14½×18in (36cm×45cm).

BACK LIGHT

When the major light source is directly behind the subject, say at dawn or dusk, that subject is thrown into near-silhouette. Forms are flattened and details are vague, and the overall impression is of low-key, shadowy tones and cool colours. Back light is moody and atmospheric, creating a sense of mystery and stillness, as in Doug Dawson's *Study of Sunset* (above). Sometimes the light catches the edges of objects, producing a lovely 'halo' of light; you can see this on the backs of the cows in this pastoral scene. To capture the effect of backlighting, work mainly within a narrow range of mid-to-dark tones; keep colours cool and muted; downplay details; and brighten the edges of forms where appropriate to indicate the way light glances off them.

◁ **The Park, Rethymnon, Crete** by Andrew Macara, oil on canvas, 10in×12in (25cm×30cm).

FRONT LIGHT

When the source of light comes from directly behind the painter and the subject is lit from the front, descriptive modeling is reduced to a minimum because there is little or no shadow. However, this is more than compensated for by the rich, decorative tones and colours which front light brings out. If shadows that are cast on the ground from behind the painter are allowed to intrude into the picture, a striking composition of light and dark patterns can result. Andrew Macara exploits this to the full in *The Park, Rethymnon, Crete* (above), devoting the large expanse of the foreground to the exploration of hard and soft shadows cast by sunlight shining from behind him. These shadows also help to draw the viewer's eye into the picture.

◁ **Moulin Huet** by Lucy Willis, watercolour, 15in×11in (38cm×28cm). Backlighting produces bright rims of light around the figures in this painting and makes them stand out clearly against the dark rocks behind. The sense of depth is increased, too, as the distant headland dissolves into a misty haze.

▷ **Evening, Sennen** by Ken Howard, oil on board, 12in×8in (30cm×20cm). Here the figures are silhouetted against the light, reflective surface of the sea. The figures are summed up with great economy using quick strokes.

▽ **South Beach, Scarborough** by Andrew Macara, oil on canvas, 12in×14in (30cm×35cm). The artist's use of backlighting and long shadows creates a marvellous evocation of a late afternoon at the beach, when the light is just beginning to turn cool.

Fog And Mist

Sunny, warm days are much more inviting to the painter than wet, cold days. Yet if the practical problems of painting in bad weather can be overcome, painting in bad weather is exciting because of the transformation that takes place in the landscape. Rainstorms, fog, ice, and snow: each of these weather conditions has its own magic, and can turn an ordinary scene into a memorable painting.

On a foggy day you will notice how contrasts in colour and tone are reduced to subtle modulations. The water vapour held in the air acts like a filter, neutralizing bright colours and flattening masses and contours. Dark tones appear lighter than normal, and light tones appear darker. The mist disperses the light so that shadows disappear and tonal contrasts are minimized. The effects of aerial perspective are more pronounced, so that objects only a short distance away seem to melt into a blur.

Capturing the veiled and mysterious aspect of a foggy landscape requires sensitive observation and a certain amount of restraint in handling the paint. Use subdued, harmonious colours and closely-related tones; too many sharp contrasts will destroy the illusion of all-enveloping mist. In watercolour, washes can be applied to damp paper and allowed to blur at the edges. In acrylics, oils and pastels, use the techniques of blending and scumbling. Reserve any clear details and dark tones for the foreground; these will enhance, through contrast, the mistiness in the background.

△ **Alsmeer Evening** by Derek Mynott, oil on canvas, 24in×36in (60cm×90cm). Here the setting sun glimmers through the evening haze, suffusing the cool blues of sky and water with its warmth. Mynott has created a symphony of harmonious high-key colours which capture the tranquility of the scene.

▷ **Early Morning, River Gade** by Brian Bennett, oil on canvas, 40in×30in (100cm×75cm).
In this high-key painting the artist has used the slow-drying properties of oils to full advantage, blending the forms into the surrounding atmosphere to give an impression of heavy mist on the river.

◁ **Winter Morning** by Edward Seago (1910-1974). Oil on board, 7¾in×9¾in (19cm×25cm).
The artist has captured the misty chill of a winter morning with vigorous brush-work and an economical description of both cattle and landscape. The muted colour scheme of blues and greys is enlivened by the flecks of warm sunlight glancing off the backs of the cattle.

STORMS AND RAIN

Some of the most atmospheric and dramatic effects of light occur just before, during, or immediately after a downpour of rain. It is possible to represent rain itself in a graphic or symbolic manner by manipulating the paint. This is particularly easy in a fluid medium like watercolour; try tilting your board downwards so that the sky washes run towards the horizon. When the washes are dry, use spattering and drybrush strokes to indicate windblown rain. In both oils and watercolour you can scratch into the paint with a sharp point to indicate slanting rain. Don't overdo these effects though, as the painting may look slick and mannered; just the merest indication of rain is all the viewer needs.

It is often more appropriate to paint not the rain itself, but some of the effects it can have. For instance, moisture in the air has a flattening effect on forms, and the effects of aerial perspective are evident over surprisingly short distances. Indicate this by painting features in the middle ground with pale, hazy tones and blending them into the surrounding atmosphere, while reducing distant features to a blur. Just after a rain shower, the sky brightens and flat, wet surfaces will act as mirrors, reflecting the light. Indicate wet pavements and rooftops with pale, wet-in-wet washes and then overlay contrasting dark reflections when the under-washes are dry.

The approach of a storm, of course, offers great potential for dramatic effects. Suggest the brooding, threatening atmosphere by painting the land and the upper part of the sky with dark tones, leaving a strip of pale sky near the horizon.

Painting rain presents practical difficulties for the artist, but it is well worth taking up the challenge. Set up your easel in a spot where you can quickly find shelter in showery weather, or retreat to the warmth and comfort of your car. I started my watercolour painting, *Rainy Day, Oporto* in what I thought were fair conditions, sitting on the bank of the river. I chose my subject because I was attracted by the boat, its reflection in the water, and the town heaped up on a hill in the

background. I was somewhat daunted, though, by the endless number of doors and windows, receding into the distance, which I would have to tackle. I started by painting the boat in strong, dark washes, but before these had a chance to dry the sky suddenly blackened and the rain fell in enormous drops all over my paper.

I dashed for the car in a confusion of paints, paper and water. Just as I got settled and was about to resume work, I knocked over my jar of water, got even more drenched, and had to set out again to refill my jar from the river. When I finally organized myself I began to wonder if it was, in fact, time to pack it in.

Having done the best I could to mop up the black smudges caused by the rain, I looked up and noticed that the windows of the car had all steamed up. Suddenly the view of the town was much more manageable! The mass of doors and windows had been simplified for me, and I had no hesitation in suggesting the distant parts of the town in one or two washes. When I wanted to introduce a bit more detail into the foreground areas in order to bring them closer, I simply wiped the windscreen with my hand.

▷ **After the Rain** by Lucy Willis, watercolour, 30in×22in (75cm×55cm). Here I have made use of light reflected on the wet ground to recreate that fresh, clear atmosphere which occurs after a rainstorm.

◁ **Rainy Day, Oporto** by Lucy Willis, watercolour, 15in×22in (38cm×55cm). Painting in the rain: though skies may be stormy and grey, interesting effects of light can often be seen.

△ **Walpole Bay, Margate** by Pamela
Kay, oil on board, 8in×11in
(20cm×27cm).
The artist worked quickly, using
directional brushstrokes, to create this
vigorous image of a blustery day. Note
the variety of colour in the sky, ranging
from white, through yellows, pinks and
blues to deepest indigo.

△ **Southwark Bridge with Cannon Street Station** by Jacqueline Rizvi, watercolour and body colour, 8¾in × 12in (22cm × 30cm).
The forms of the bridge and buildings almost dissolve into the surrounding atmosphere in this impressionistic painting. Darker values and animated shapes in the foreground sky give an impression of scudding clouds.

SNOW SCENES

Having struggled through the summer months with green, which is the predominant and, arguably, the most difficult colour in many landscapes, painting snow scenes comes as a welcome change. The interesting thing about snow scenes is that the landscape appears much more tonal. On an overcast day, the snow-covered land is, for once, lighter in tone than the sky, and the dark shapes of buildings and trees stand out in sharp contrast with the lighter tone of the snow. This can make for some very dramatic and exciting compositions, with a cool, moody atmosphere.

In complete contrast, sunny-day snow scenes present an exhilarating, high-key mood. The snow becomes a great sheet of white paper on which blue-violet shadows and golden highlights are drawn. You will notice this particularly in the early morning and late afternoon, when the upright snow-covered planes reflect the warm yellow of the sunlight, while the shadows and dips and hollows reflect the cool blue of the sky. In fact, snow is hardly ever pure white all over.

Watercolour is the most economical medium for painting snow scenes, as you can leave some areas of white paper to represent the lightest lights. Model the contours in the snow with pale washes of blues, greys and yellows. Use straight lines and hard edges where the snow is fresh and clean;

indicate patches of melting snow by using scrubby drybrush strokes to represent the earth peeping through.

When painting outdoors in very low temperature, you may find that your washes will actually freeze as you apply them. Though this can sometimes result in attractive 'snow crystal' effects, it very often means that you are pushing lumps of slush across the paper. To overcome this problem, make sure you have some gin or vodka with you. These are colourless alcohols and when added to your water act as an anti-freeze.

Any alcohol left over from this may be used to warm and enliven the artist, of course…

When painting in oils such interesting solutions are unnecessary because, of course, the paint will not freeze. It's a good idea to work on a pre-toned canvas, in a warm earth colour such as yellow ochre, and allow the colour to show through occasionally. Use a thicker impasto in the light-struck areas, smoothing the paint so it catches the light and appears more luminous.

◁ **Winter Park, Moonlight** by Robert Buhler, oil on canvas, 20in×24in (50cm×60cm).
In this low-key painting, Robert Buhler has caught the still, chill atmosphere of dusk on a snowy winter's evening. The closely-related tones, muted colours and considered composition all combine to emphasize the mood of the picture.

△ **Frosty Morning on the Road to Parc** by Diana Armfield, oil on board, 21¼×22in (53cm×55cm).
The frost on a crisp morning in winter transforms a country lane and presents an exciting subject for the exploration of colour. The artist has used a warm beige underpainting which vibrates with the cool colours in the shadows. The combination of these delicate hues gives a pearly irridescence to the landscape.

LIGHT IN SKIES

The great theatrical effects of light and colour in the sky have produced some of the most magnificent and sublime works in the history of art. Glorious sunsets, towering banks of sunlit cumulus cloud, dark, storm-filled skies, the feathery patterns of cirrus cloud against a clear blue expanse: these are the kind of subjects that have attracted artists for generations.

The great English landscape painter John Constable declared that "The sky is the source of light in nature and governs everything." Thus, the quality of the light in the sky will have a direct influence on the atmosphere and mood in your landscape paintings. Or, at least, it should. Inexperienced painters sometimes make the mistake of treating sky and land as two completely separate entities, resulting in a disjointed picture with no real feeling of light. One way to overcome this problem is to bring the sky and the landscape along simultaneously, working from one to the other and bringing some of the sky colour into the land and vice versa.

It's also important to compare the tonal relationship between the clouds and the sky, and between the sky and the landscape. For example, the tones in the sky are almost always lighter than those in the land. If you see a dark cloud, rimmed with light, this may look at first glance like the contrast between black and white. If you then look at the branches of a tree in dark silhouette in front of that cloud, you will see that the dark of the cloud suddenly appears, in this context, as a middle grey. Through this delicate balancing of tones you can capture the brilliance of the sky without it appearing to be an unnatural, conflicting and separate entity within your composition. By relying on observation and analysis of this kind you can paint the most extraordinary skies and find that they can be carried off with conviction, even if they appeared utterly improbable at first.

◁ **Sunrise, Nine Springs** by Ronald Jesty, watercolour, 8in×12in (20cm×30cm).
Ronald Jesty has captured one of those beautiful, accidental phenomena which make the sky a constant source of wonder and inspiration. The dark silhouette of the landscape accentuates the brilliance of the sky and focuses our attention upon the wonderful conformity of the clouds.

△ **Wartime** by Albert Goodwin (1845-1932). Oil on canvas, 14in×20½in (35cm×51cm).
Albert Goodwin conveys a powerful feeling of perspective and scale in this painting. See what a large expanse of the canvas is devoted to the broad mass of clouds which scud overhead and draw us into the picture.

SELECT AND SIMPLIFY

Very often when painting skies you will be painting from memory, as everything moves and changes so quickly. You must develop a facility for assessing colour, tone, volume and shape quickly and accurately, something which can only be achieved through constant practice. Before you begin painting, spend some time just watching the changing patterns above you. Look, too, at the way in which the sun illuminates the clouds. It may seem chaotic at first, but light has a simple logic when you study it more closely. You will notice, for instance, that the sky is usually lighter near the sun and grows increasingly darker as you look away from the sun. On a clear day the sky directly above is a brilliant blue, which grows lighter and cooler towards the horizon. This colour change is noticeable even if the sky is cluttered with clouds. Remember, too, that aerial perspective acts on clouds just as it does on land; white clouds directly above will contrast more strongly with the darker tone of the sky than similar clouds do in the distance.

Clouds are rounded forms, so model them as you would any spherical shape. There will be one main highlight, reflecting light from the sun, while the bottom and sides will reflect colour and light from adjacent clouds, the sky, and the earth below. Give the clouds roundness and atmosphere by blending and softening most of their edges wet-in-wet, or by scumbling with thin, dry colour.

△ **A Moonlit Road** by William
Lionel Wyllie, watercolour, 4½in×2½in
(11cm×6cm).

△ **Moonlight, Norway** by William
Lionel Wyllie, watercolour, 13in×9in
(32cm×22cm).

BE BOLD

As your painting develops you may be tempted to capture every change and nuance in the sky by repainting again and again. Overworking soon kills the translucency of the sky, however, so resolve to be bold and direct in your execution. In oils and acrylics you can scrub out passages or change them by overpainting if necessary, but try to keep the brushwork looking spontaneous and free. Notice how, in many great paintings of skies, the artist has used much freer, broader brushstrokes towards the top of the picture, giving the sky a strong sense of scale and perspective.

In watercolour there is a risk of spoiling a successful landscape painting by adding the sky in afterwards. If your courage fails, you might opt for a non-commital sky which, while it won't ruin the picture, certainly won't enhance it. It is better to introduce the sky at an early stage so that the tonal relationships can be built up together. When practising painting skies in watercolour, make large numbers of quick studies, and be prepared to discard readily anything that goes wrong. Do not be afraid of making mistakes, but profit by them and apply what you have learned to the next painting.

The sky is a constant source of wonder and inspiration. The more you observe it with a painter's eye, the more you will understand its logic, and the greater will be your success in translating these observations onto canvas or paper.

△ **Moonlit Mackerel Sky, France** by William Lionel Wyllie, watercolour, 10in×6in (25cm×15cm).
By making numerous, quick studies of skies, like these ones by William Lionel Wyllie (1851-1931), you will develop your powers of observation and analysis, as well as the technical skill required to describe the effects of light in the sky.

Demonstration: Light Weights

Peter Allen Nisbet has been fascinated by skies all his life, and now spends much of his time painting them. Here he describes his method of painting clouds in oils, using a glazing technique in which layers of thinly-applied colour interact to create the illusion of dense air.

1

1 I always begin a painting by making a series of charcoal sketches, which helps me to establish the tones. Here I was interested in the huge cloud hovering over a vast expanse of water, backlit by the morning sun.

2

2 I now make a light outline sketch on the canvas, using conté pencil. I apply an underpainting over the sketch to establish values and set a warm undertone that will influence later colours. I use raw sienna mixed with Winsor & Newton underpainting white, and touches of Winsor violet and ultramarine. I dilute the colours with five parts rectified turpentine to one part damar varnish, and apply them loosely to the canvas like watercolour.

3

3 Beginning with the horizon, I build up the sky colours with successive glazes. The yellows consist of cadmium yellow pale, Winsor lemon and yellow ochre, mixed with zinc white or titanium white in the lightest areas. The blues and greys are mixed from ultramarine blue, cerulean blue, Winsor violet and rose doré. My medium consists of five parts turpentine, one part damar and one part stand oil. Each successive layer of paint receives slightly more oil; this 'fat-over-lean' rule helps to prevent the paint surface from cracking as it dries.

4 I continue adding more glazes, and use a sable fan blender to soften the forms of the large cloud. The paint is applied thinly to allow the underpainting tones to influence the overall colour scheme of the cloud. I finish off by reinforcing the main lights and darks, and lessen the intensity of the shadow rays.

▽ This detail from the finished painting shows how the cloud tops reflect the blue colour of the sky, while the bottoms absorb and reflect the warmer earth tones. I built up the three-dimensional form of the main cloud with many thin layers of paint. In contrast, the clouds near the horizon were painted with just one or two horizontal strokes.

4

△ **Light Weights** by P. A. Nisbet, oil on canvas, 14in × 24in (35cm × 60cm).

LIGHT ON WATER

The surface of a large expanse of water is in a sense like a huge mirror reflecting the light of the sky. When the surface is calm it can reflect the sky almost exactly; when the surface is ruffled by a breeze the reflected image of the sky is broken into tiny fragments. As the water moves, the patterns of lights shining on and through it are constantly broken and reassembled, making it difficult to analyze exactly what you are seeing. It is essential, therefore, to study at length the repeated patterns of light and colour before committing yourself to your painting. The difficulty lies in creating a clear image of what is happening. Having looked closely at the water, close your eyes and try to visualize it in your mind's eye. Notice, in your mental picture, where your visualization is weak or vague. Open your eyes and fill in the missing details. Repeat this process a few times until this mental picture has become absolutely clear, and only then start your painting. The ancient Japanese painters under-stood this process entirely, and would sit looking at their subject for hours, even days, before going away and painting – in a few swift brushstrokes – the fully realized distillation of their lengthy observations. Turner, in a rather more dramatic fashion, had himself lashed to the mast of a ship in order to observe, as closely as possible, the effects of a storm on sea and sky, which he later painted from memory.

The smooth, glassy, light-reflecting surface of water is best expressed as simply as possible. Seek out the major shapes of light and dark and omit all superfluous details; water looks wetter when painted simply. Use bold, calligraphic strokes to strike in the wavering patterns on the water's surface. Just as with painting skies, it is important to constantly assess and compare the relative tonal values, both within the various parts of the water itself and between the water, the sky, and the land. It is, after all, the contrast between light and dark values which conveys the impression of light in any picture.

▷ **Pinhao Rapids** by Lucy Willis, watercolour, 15in×11in (37cm×27cm). I was attracted to this scene by the contrast between the smooth glassiness of the water in the foreground as it slid over the boulders beneath, and the turbulent water beyond. I found that the effect of the speed and undulation of the water could be described solely through the pattern of light and dark reflections. The curious thing about water such as this, though, is that although it is moving at such high speed the reflections stay relatively still.

▽ **Blue Boat** by Ronald Jesty, watercolour, 4in×9½in (10cm×24cm). Here the artist has used just four simple tones to succinctly suggest ripples in still water.

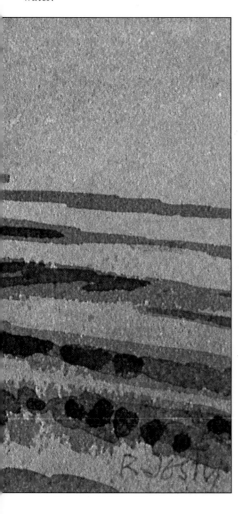

◁ **September Floods Near Glastonbury** by Arthur Maderson, oil on canvas, 48in×34in (120cm×85cm). Notice the unity of brushwork in this painting; the trees, the sky and the reflections in the water are all handled in exactly the same way. Maderson has not painted the water by trying to actually paint water; he has simply created the illusion of water, tricked our eyes into believing it is there, by painting the reflections in it.

△ **The Drowning Pool** by Nigel Casseldine, oil on gesso, 20in×24in (50cm×60cm).
By analyzing the patterns of light and dark tone and warm and cool colour, the artist has skilfully defined both the surface of the pond and the apparent depth of the water. The matt crust of pond weed emphasizes the glassiness of the reflection in the water and appears to float above it.

Demonstration: View of Venice

In this demonstration, artist John Martin describes his method of painting water, simply and with a minimum of fuss. He begins with broad washes of transparent watercolour, then overlays these with mixtures of watercolour and gouache for the highlights and reflections. Although gouache is opaque, Martin finds that its brilliant, light-reflective qualities are perfect for capturing the sparkling quality of the atmosphere surrounding a body of water.

1

2

1 I begin by plotting the main shapes of the composition in pencil. Then, using a soft, round brush, I apply pale underwashes of blue and grey to establish the overall colour key of the painting. For the sky I mix up varying proportions of permanent white, cobalt blue and ultramarine, and apply the colour with loose, scumbled strokes that give a sense of movement.

2 I add further layers of colour and continue to build up the main areas and masses. Using a mixture of ultramarine, vermilion and white, I make a warm grey, which I use to strike in the shapes of the doors and windows. With the same colour, I paint the wavering reflections in the water, using the pointed tip of the brush. Then I roughly indicate the shapes of the boats and their reflections, using cadmium yellow, cobalt blue, ultramarine, burnt umber and viridian.

3

4

View of Venice by John Martin, watercolour and gouache, 6in×6in (15cm×15cm).

3 Having established the cool blues, I now turn my attention to the warm yellows in the foreground. I indicate reflected sunlight on the nearer buildings with mixtures of lemon yellow, alizarin crimson and white. Then I add thin washes of yellow in the water. When these are dry I strengthen the details of the boats and reflections, using horizontal and vertical strokes of blue, grey and white.

4 In the final stage I strengthen the darks and touch in the highlights: white on the roofline of the distant buildings, white and lemon yellow on the nearer buildings, and mixtures of white and cobalt blue on the water. In the near foreground I use thicker paint, applied with swift drybrush strokes, to indicate sparkling light on the water's surface.

This detail of the water shows how I use a small, just-damp brush and thickish paint to make dry, broken strokes that give a sense of movement and light on the water's surface.

CITYSCAPES

The emphasis so far in this chapter has been on the great natural settings of land, sea and sky. But most of us nowadays live in towns and cities – where there is just as much to inspire the painter with imagination and an enquiring eye. Some of the lovelier cities – Paris, Rome, Venice – have of course attracted painters for generations. But many recent artists have revealed just how beautiful and moving even the most ordinary modern urban environment can be. The American Edward Hopper captured the lonely desolation of empty sidestreets and the neon glare of seedy cafés; L.S. Lowry took his inspiration from the mean streets of England's industrial north.

Cityscapes provide an excellent opportunity to study the play of light on shapes and surfaces. Observe the changes that take place in the city as the day progresses: from the hazy blue silhouettes of buildings in the early morning light, to the sharp, angular shadows and reflections under the midday sun, to the jumbled skyline etched in black against the evening sky. And don't forget the night-time world, which does not fade into darkness but comes brilliantly alive with a whole spectrum of artificial light.

Inclement weather, too, can invest a city scene with atmosphere and drama. Think of a dark winter's day, heavy with fog, when the gleam from office windows and car headlights shines in the gloom, or the way rooftops and pavements sparkle after a shower of rain. The magic comes not so much from the buildings themselves, but from the way light falls on them. Even a drab old brick wall comes alive when golden afternoon sunlight glances across it, or when the long, slanting shadows of a nearby building, or balcony, or fire escape, create graphic patterns across its surface.

It is, of course, impossible to set up your easel in a city street and paint on-the-spot, unless you wish to attract a crowd of curious onlookers. It is here that the training of your visual memory, aided by sketches and photographs if you like, becomes vitally important. Jot down notes on any scrap of paper as you pass, to record any particular effects of colour and light that have struck you. These can be enormously valuable when you come to recreate the scene later in your studio or home.

◁ **China Town** by Doug Dawson, pastel, 20in×22in (50cm×55cm). The vibrant colours in this pastel stand out brilliantly against the surrounding dark. This is a strongly suggestive image evoking all the excitement and mystery of the city at night.

▷ **Quadri's Band, Piazza San Marco** by Diana Armfield, pastel, 10in×8in (25cm×20cm).

That moment at dusk when the lights of the city take over from daylight has been beautifully captured in pastel by Diana Armfield. The neutral colour of the paper shows through and sets off both the cool colours of the sky and the yellow glow of the electric lights.

▽ **The Sunken Tree, Strand on the Green** by William Bowyer, oil on board, 40in×50in (102cm×127cm).

In the late afternoon, sunlight slants low across the landscape and creates dramatic contrasts of light and shade. Here the cool, dark colours of the tree and the water lend emphasis to the brilliant light striking the buildings.

Project: night and day

City streets are a fascinating subject for the study of light. Because the buildings are so closely confined, shadows fall from one to another, creating striking patterns of light and shade. For this project, find a convenient vantage point such as an upper-storey window which gives you a clear view of a city street. Make quick studies of the street, in any medium you like, at four different times of the day: morning, midday, late afternoon and night. Your completed studies should provide a record of the changing temperature of the

1

1 Morning Early in the morning, the scene was low-key with cool colours. The street was still in shadow, so there was little contrast in either colour or tone except for the sunlit corner of the far building and the glow of the two hazard lamps in the foreground, where workmen had been digging up the road.

2 Midday Returning at midday from a shopping trip, I found that the sun had moved round but the street was still surprisingly shadowy. The sky was now a brilliant blue and the sun was beating strongly. Although still predominantly low-key, the colours were becoming warmer and the tonal contrasts stronger.

2

light as the day progresses, as well as the way shadows move and lengthen.

I painted the watercolour sketches shown here while on holiday in Italy. I enjoyed leaning precariously out of the window of my apartment, making studies of the quiet street below. I worked in watercolour, using a limited palette consisting of cobalt blue, Indian red, cadmium yellow and Vandyke brown, mixing black and brown Indian ink with the paint in the shadow areas to give them more body.

3

3 Late afternoon Returning to my vantage point in the late afternoon, I found that at last there was some sunlight shining into the street. Though there was still a cool feeling in the shadows, the sunlight was reflected back onto the buildings on the left, clarifying the detail and warming the colours.

4 Night By the time I came back from the restaurant that night the scene was transformed. Many of the windows which had been dark and shuttered during the day were now lit up, giving evidence of life and activity within. Slivers of light shone evocatively from open doorways. The central lamps, which had played a minor role in the daytime pictures, now became a focal point, while the far end of the street was cast into complete darkness.

4 Studies of a Street in Trastevere by Lucy Willis, watercolour 10in×8in (25cm×20cm)

CHAPTER SIX

PAINTING LIGHT INDOORS

Each time you open a curtain, close a door, or switch on a lamp, you create a different world of light and space. Whether you are painting a portrait, a still life, or the room itself, the type of lighting you choose will radically affect the emphasis and atmosphere. It is light which controls the emotional impact of a painting and gives shape, life and meaning to the subject.

This chapter looks at the many ways in which you can control the light, experiment with it creatively and use it to enhance the particular qualities that you wish your picture to convey.

Controlling The Light

Light is something you really can control when painting indoors. Before you begin a painting, first consider what kind of lighting you have available, what effects you can achieve with it, and what kind of mood or atmosphere you wish to evoke in your finished picture. It is well worth experimenting with a variety of light sources and lighting set-ups in order to familiarize yourself with the creative possibilities of light.

Artificial Light

Most artists prefer to paint by natural daylight, since artificial light tends to distort colours somewhat. However, the advantage of artificial light is that it is both constant and controllable. If you do work with artificial light, you might like to try one of the modern fluorescent 'daylight' bulbs, which is the closest you'll get to the real thing.

When illuminating your subject with artificial light, don't forget that the light source itself can be included as part of the composition. For example, the warm pool of light cast by a table lamp or floor lamp in a darkened room can be particularly attractive. Be aware, too, that objects close to the main pool of light will be affected by the diffuse light that glows through the lampshade. This can colour other elements in the composition with the colour of the lampshade itself, thus effecting a harmonizing influence on the composition and helping to tie the picture together. Table lamps and floor lamps can create a cosy and intimate atmosphere,

radiating a soft, warm glow throughout the picture.

You might, on the other hand, prefer to use strong, directional light which illuminates the subject from one side, casting powerful, dark shadows which themselves become essential elements within the composition. Spotlights and anglepoise desk lamps are particularly suitable for this, as they can be angled in whatever direction you like. If you feel that the shadows cast in this way are too uniformly dark, you can effectively add some reflected light to the dark shadowy area by setting up a white or light coloured board opposite the source of light. This will bounce light back onto the objects in the composition, gently illuminating areas of shadow and also adding to them a soft glow of colour if you have chosen to use a coloured board.

You can make more complex patterns of light and shadow by introducing a second or even third spotlight and shining them onto the subject from different directions. The pattern of overlapping shadows which this creates can be intriguing and dramatic. However, it's important to be aware that the more sources of light you introduce, the more the tonal contrasts and the modelling of form will be diminished.

This is also the case with overhead ceiling lights and especially fluorescent strip lighting. Though this bright, diffuse light creates a particular kind of atmosphere which might be appropriate in certain situations, it can equally have a deadening effect on even the most interesting set-up. Cast shadows are reduced to a minimum, and so contrasts

between light and dark areas play a secondary role.

NATURAL LIGHT

In the northern hemisphere, painters working indoors often prefer to work with light from a north-facing window, which has the advantage of being cool, clear and constant. Unlike light from other directions, north light does not cast heavy, changing shadows, and therefore enables the painter to work throughout the day without worrying about the position of the sun. This can be extremely helpful if, for example, you want to work continuously on a portrait or a complex still life which may take a number of days to complete.

On occasions, however, strong, direct light entering a room through a south-, east- or west-facing window will provide the inspiration for a painting. For example, the shadows cast by a window frame into a sunny interior, or across a figure, can be particularly evocative. Because these shadows will move, just as they do when you are painting outdoors, you will have to work quickly.

When painting by natural daylight from a window, you cannot move the source of light, but you can modify it. Consider using the light filtered through a lace curtain, through partially opened shutters, or, more dramatically, through a venetian blind. In each case a completely different atmosphere is created. Consider, too, the position of your subject relative to the light source: do you want your still life, or your sitter, to be lit from the side, the back or the front? It's worth spending some time positioning and repositioning your subject until you find the effect you're seeking. At the same time, try to ensure that your easel is positioned so that the light falls over your left shoulder if you are right-handed (and vice versa) so as to avoid distracting shadows falling on your paper or canvas.

Some interesting effects can be created by combining daylight with artificial light. Electric light is warmer and more yellow than daylight, which is usually blue and cool except on very sunny days. The combination of daylight and artificial light creates a subtle play of warm and cool colours which can noticeably enrich a painting. The shadows cast by cool light from the window appear warm and glowing because they absorb the warm light from the lamp, while the shadows cast by the lamplight can appear strikingly blue or violet as they absorb the cool light from outside.

△**The White Coffee Pot** by Lucy Willis, watercolour, 20in×24in (50cm×60cm). Diffuse daylight enters the dark room from a window on the left, creating gentle blue shadows. Light reflected from surfaces in the room enriches the colours within these shadows.

▷**The Black Bowl** by Lucy Willis, watercolour, 20in×24in (50cm×60cm). On an overcast day I decided to set up a similar still life, this time introducing artificial light to brighten the composition. Cool, blue daylight complements the warm, yellow light shed by the electric lamp.

◁**Quiet Noon** by Fred Dubery, oil on board, 10in×18in (25cm×45cm). Fred Dubery has positioned his subject close to a French window, which casts interesting patterns of light and shade onto the table. These patterns, besides creating a quiet, peaceful mood, are also an integral part of the composition, encouraging the viewer's eye to move across the picture.

◁**Summer Table** by Annie Williams, watercolour, 15in×10in (37cm×25cm). There is a lively, sunny-morning atmosphere in this fresh and simple composition. Sunlight streams through the window, lightening and intensifying the colours in the room.

◁**The Floor Lamp** by Lucy Willis, watercolour, 22in×15in (55cm×37cm). This was painted as I sat on the floor late one evening. The room was lit solely by the floor lamp featured in the painting, which shed a low arc of light that created a cosy, cheerful atmosphere.

▽**Victoria with Cat** by Pamela Kay, pastel, 10½in×14½in (26cm×36cm). Here we have a clear example of how light can create mood: there is a warm, intimate atmosphere in this portrait. The artist has worked on a dark blue paper, which acts as a foil for the warm luminosity of the lamplight.

INTERIORS

As with a landscape or a cityscape, an interior can either be the setting for human activity or it can be the subject of a painting in itself. The different painting genres – still life, portrait, nude, interior – are not rigidly distinct from one another, and many paintings are combinations of some or all of these. However, the details of an interior may be most clearly appreciated where there are no other dominant themes to draw the attention. Indeed, an empty room can have a particular poignancy because of the very absence of people. The feeling of light, space and atmosphere fills the canvas, while the viewer's imagination is free to explore the details within.

On the other hand, any interior with figures in it becomes a setting for those figures and gives dimension and meaning to their activity. People can be 'interrupted' in some task which helps to define them: combing hair, chopping vegetables, eating a meal or playing music. The interior becomes a vehicle for capturing the small, intimate moments of life.

It's a good idea to make several studies of a favourite room, at different times of the day and under different weather conditions. Begin with monochrome wash studies, which will give you an impression of the overall tonality of the room. View the room from different angles, and re-arrange the objects and furniture if necessary in order to make a pleasing composition. The colours in an interior may seem more sombre than those outdoors, because the light is not so strong; but they are never dull. Even in the deepest shadows you will see rich browns and reds and cool blues and greens, particularly on a sunny day, when reflected light from the window bounces off the walls, floor and ceiling. Under diffuse light, subtle colours and gentle shifts of tone create a different, more contemplative mood.

Painting an interior gives you a choice of light source: natural, artificial, or a combination of both. Tonal values and colour temperature are major considerations when it comes to establishing the mood and ambience of a room. Don't be too literal in your approach to painting the room and the objects within it; remember that local colours are affected by the colour of the prevailing light. Generally, natural light is cool and bluish (except for direct sunlight), while artificial light is warm and yellowish. Reflect this difference in your colour mixtures, not forgetting that warm light produces cool shadows and vice versa, and you will achieve a satisfying sense of all-pervading light and atmosphere.

◁ **Sunday Lunch** by Lucy Willis, watercolour, 11in×15in (27cm×37cm). Here the interior is the setting for a family portrait: I worked frantically to catch the scene, concentrating on the way the light picked out each figure. The little chinks of white paper are evidence of haste, but I refrained from filling them in as they add a certain vivacity.

△ **Ballet Lesson** by Andrew Macara, oil on canvas, 14in×16in (35cm×40cm). This charming and unexpected composition is illuminated by large windows which cast patterns of sunlight onto the floor. These areas of warm colour emphasize the coolness of the interior and the atmosphere of quiet concentration.

▷**Autumn Window** by Jack Millar, oil on board, 35in×41in (89cm×104cm). Afternoon sunlight slants through the Venetian blinds and creates a hazy warmth in this interior. Note how the cool blues on the window frame and the backlit figure accentuate this warmth.

INSIDE LOOKING OUT

In most interior scenes, a window or an open doorway will give you the chance to add an extra dimension to your composition by providing a tantalizing glimpse of 'something beyond', which draws the viewer deeper into the picture. It is curious how the view through a window, however mundane, always looks so inviting.

In addition, of course, windows and doorways provide a ready-made 'frame' through which the artist can explore light. One of the most interesting aspects of painting from inside looking out is the way in which the mood and atmosphere of an interior can change from day to day, even hour to hour, depending on the weather and the position of the sun. In winter the sun is weak, and low in the sky, and casts long, soft shadows which evoke a melancholy or a meditative air. In summer the sun is high in the sky and casts

short, hard shadows and lovely flashes of reflected light. If a window is small, the interior may well be dim and shadowy, while outside all is bright and high-key; this contrast is irresistible, creating as it does a 'picture within a picture'.

Before I started *Door to the Garden* (opposite) I was going to paint outside in the sunshine, but I had often noticed in passing how powerfully illuminated the garden appeared when framed by the doorway and in contrast to the cool interior; I decided to make this the subject of my painting. When planning my composition I decided the doorway would appear more enticing if it was only a small area within a much larger space.

The subject of the painting was really the light: its brilliance, the different colours of the shadows it created, the reflections in the glass-covered pictures on the walls. One of

▷**Conservatory** by Jack Millar, oil on board, 30in×25in (75cm×63cm). Vibrant, high-key colours are used to great effect here in conveying the feeling of a bright, hot day. The open doors create a 'frame within a frame' which focuses our attention on the figure in the background.

◁**Door to the Garden** by Lucy Willis, watercolour, 22in×30in (55cm×75cm). I started this watercolour by painting the doorway, some of the foliage, and the pattern of light and shade on the floor. The angle of the sun began to change, so I stopped painting and resumed work at the same time the following day. I painted each of the walls as one continuous wash, cutting around the shapes of the pictures and lightening, darkening, or adding colour wet-in-wet to create the effects of reflected light or deepening shadow.

the attractions of strong sunlight shining into an interior is the way that this light can project an image in shadow of the outside world onto the interior surfaces. Here I particularly liked the way the leaf shadows from nearby trees broke into the severe rectangle of the light from the doorway.

As well as this projected shadow there are patches of reflected light, and also other, softer shadows cast by objects within the room. On the wall to the right of the doorway, for instance, you will see a slight warm glow, created by light reflected from the open door. I enjoy painting reflections in window and picture glass and there was ample opportunity to indulge myself here: I had a window behind me, the door in

front, and lots of pictures all around. I had to consider carefully the tonal values of these reflections and subdue those on the left so they would not detract from the brilliance of the light through the doorway.

Painting glass, which some people find mysterious, is relatively easy as long as something is reflected in it. Then it becomes simply a matter of half-closing your eyes and analyzing exactly what you see in terms of tonal values. Where there are strong reflections like those here, everything behind the glass becomes relatively dark.

Project: atmosphere and space

Choose a favourite room and select a view of it which
incorporates a window facing you. Make three studies of
the room from the same point of view, first in daylight,
then at dusk, and then again at night. Use any medium
you like. The object is to see how different light
conditions affect the atmosphere inside the room and
alter the impression of space. Observe the change in
tonal relationships as the day progresses: the window is
lighter than the interior in daytime, closer in tone at
dusk, and darker at night.

1 Daytime You will notice that even on a dull day the tonal
value of the window is much brighter than anything within the
room. Assessing the tone of the wall surrounding the window
can be a problem, especially if the walls are white. Don't let
your knowledge that the wall is white interfere with your
perception; if it is in deep shadow, paint it with strong dark
tones.

2 Dusk Return to your painting position at dusk and turn on
an overhead electric light. Try to capture that particular
moment when the tone of the evening light outside the window
is the same as that of the window frame and its immediate
surroundings. There is at this point an especially exciting
colour relationship between the cool blue light outside and the
warm yellowish electric light inside. The tones are roughly the
same but the colours are vibrantly different.

The Arched Window by Lucy Willis, pastel, 18in×12in (45cm×30cm).

3 Night This time illuminate the room with one or two table or floor lamps, making sure there is enough light for you to work by. There will be areas of deep shadow and areas of intense luminosity. Look for subtle variations of warm and cool colour within the shadows and paint them with swift, flowing strokes, merging colours together where soft transitions of colour or tone occur. Make sure the shadows are deep enough to make the warm, glowing light from the lamps really stand out.

Now the tonal relationship between the window and the room will be a complete reversal of that in the first painting in the exercise. Depending upon the position of the lamp, you might see some reflections in the glass, in which case half-close your eyes and establish the tonal value and the shape of the light reflection. Try to represent it with only a few brief dabs of the appropriate light colour, rather than focusing in on the reflected image and breaking up the flowing quality of the painting.

STILL LIFE

The roots of still-life painting go back to the beginning of the 16th century, when artists began to devote whole canvases to subjects which had, until then, been no more than details in a larger composition: the food on a table, the flowers in a vase, a selection of kitchen objects. Since then, of course, the still life has become a genre in its own right.

Although the traditional idea of a still life is that its components have been deliberately placed and the lighting carefully arranged, it is equally possible to find the inspiration for a still-life painting in some entirely accidental arrangement of objects. The happiest compositions can occur without any deliberate intervention on the part of the artist; it may be a few breakfast things left on the kitchen table, or even the children's toys heaped in a corner of the room. Often, there is something about the play of light and shadow which draws you to a particular arrangement of shapes; the most ordinary objects, once you begin to see them in terms of light, can be an inspiration.

When choosing objects for a still-life group, it is advisable, as always, to start simple. Don't feel obliged to load your kitchen table with artichokes, bottles of wine, dead pheasants and plaited loaves of bread. Why not take one or two simple objects to begin with – some fruit, a bowl and a bottle, or an onion and a carrot – and study them in depth? There is plenty to explore, in terms of light, colour and form, in even the simplest and most unassuming vegetable.

LIGHT AND MOOD

Whether you choose to illuminate your still life with natural daylight or artificial light is a matter of preference. The main thing is to arrange the lighting to full advantage; that is, to use it as a tool, not only to express form through the contrast of

light and shade, but also to establish a particular mood. Remember, you are not simply painting a group of objects: you are painting a group of objects *suffused with light*.

One way to quickly familiarize yourself with lighting effects and the moods they convey is to set up a simple still-life group and paint it under a range of different lighting conditions. Place the objects on a tray, which will enable you to move your still life from one part of the room to another, should you wish to, without disturbing the relationship of the objects. Study the same group under both natural and artificial light, and a combination of both. Observe the effects of bright sunlight, diffuse light, three-quarter light, back-light, side light, and so on. You will soon become aware of the different moods and effects created by altering the strength and direction of the light, and how the impression of three-dimensional form is either heightened or lessened. Here are just a few of the options available:

Back light Place your tray of objects on a table directly in front of a window. Position yourself so that you face the window with the tray in front of you. This arrangement, in which the subject is placed between you and the source of light, is sometimes known by the French term *contre jour* ('against the light'). Lit from behind, the objects are seen in near-silhouette, and the overall effect is of soft tones and muted colours which create a soft, romantic mood.

Side light Place a desk lamp to one side of your still-life group and shine it onto the objects from a low angle, say a foot above the table. You will see that each object is clearly divided into a bright side and a shadow side and there will be long, sharply defined cast shadows. The modelling of the forms is now much stronger. Because the light falls from a single, isolated source, everything that lies outside the beam of light is cast into shadow. This heightens the sense of drama and intensifies the mood – an effect known as 'chiaroscuro'.

Combined light sources For this third experiment in lighting, pick up your tray and move it through 45 degrees so that the still life is lit by daylight coming through the window from one side. Switch on your desk lamp and shine it onto the still life from the other side. The combination of warm electric light and cool daylight will have a magical effect on the colours of the shadows, with warm oranges and yellows combining with cool blues and violets. As the two sources of light interact, a complex pattern of hard and soft shadows is created. This painting will be the most colourful and light-filled of the three and will convey a corresponding bright and cheerful mood.

TONAL CONTRASTS

One thing to watch out for when portraying a particular lighting condition is the degree of tonal contrast between the lit areas and the shadow areas. In bright sunlight, for example, the tonal contrast between the light and shadow areas on any given object will be quite marked; in diffuse or overcast light, the tones will be much closer because of the weaker illumination. Try to keep the degree of tonal contrast consistent throughout your painting, otherwise the impression of light will be lost.

▷**The Minton Set Against the Light** by Jacqueline Rizvi, watercolour and body colour, 14in×16½in (35cm×41cm). These pieces of old Minton chinaware are illuminated by soft, cool light coming from a window behind. The painting has a simplicity and repose which is enhanced by the gently muted colours and soft tones.

◁**Summer Flowers with Cups** by Pamela Kay, oil on board, 10in×12in (25cm×30cm). Here a small, concentrated light source is positioned above the still life and to one side, so that light falls on it from a sharp angle. This creates a strong chiaroscuro effect, in which the bright flowers and glistening china are revealed in dramatic contrast against the dark, shadowy background.

▷**Bedroom Interior with White Lilac**
(detail) by Jacqueline Rizvi, watercolour
and body colour, 16in×14in
(40cm×35cm).
A still life group doesn't have to be
formally arranged; sometimes a corner of
a room, catching the light in a certain
way, will provide the impetus for an
unusual composition. This interior is
suffused by light from a nearby window,
yielding delicate shadows and subtle
colours. The still life and setting are
tenderly described in terms of closely-
relating greys.

▽**Still Life with Oil Lamp** by Ronald
Jesty, watercolour, 14in×30in
(36cm×76cm).
This unusual study has been painted
under bright artificial light. The shadows
created by the gentle, rosy light from the
oil lamp are subtle and delicate.

△**Straw Hats** by Lucy Willis, watercolour, 15in×11in (37cm×27cm). In this composition I was interested in the way the long, slanting patterns of light and shadow passed across the floor and over the hats and cushions. There is an interesting duality between the solid forms of the objects and the abstract patterns of light that play upon them.

Demonstration: Geranium Cuttings

In this demonstration I have chosen to paint a still life which is illuminated from behind. The set-up is very much as I found it, on a bench in my greenhouse, although I have moved the two nearest pots forward onto the table to make a less strictly formal composition. The light outside is bright, but overcast, so most of the shadows are soft and graduated.

The flowerpots are contrastingly dark, and it is this relationship between light and dark tones that is my main concern from the outset.

1 I have decided to use pastel for this still life, as it is an appropriate medium for describing the matt surfaces of the earthenware flowerpots and the soft shadows. I am interested in the way each object is rimmed with light, creating a repeating motif of pale elipses. As I am working on almost-white paper, I do not draw in these rims, but let them gradually emerge as the surrounding darks are built up around them with loose, vigorous, diagonal strokes. I begin by positioning the central group of shapes and establishing the dark, light, and middle tones. I use black, grey, and a couple of browns, with the occasional patch of purple, green, or blue.

2 Among the predominant neutral greys and browns there are a few isolated patches of intense colour where the light shines through translucent objects such as the orange of the plastic flowerpot and the geranium leaves. I use pure, bright colours for these, while building up the muted colours surrounding them with as many as six different neutral colours worked in together.

3 The effect of light on the window sill and table becomes much stronger now as I tone down the scene outside the window with muted greens and greys. In the lightest areas of the picture I use bold scribbles of white pastel to lift the tone and break up the expanse of bare paper. The volumes and shadows become more resolved, but I still try to keep a fairly loose approach, applying the final dashes

△**Geranium Cuttings** by Lucy Willis, pastel, 15in×22in (37cm×55cm).

of grey, brown, purple or blue in consistently brisk strokes. To pull the whole picture together I gently rub my finger over certain areas, solidifying the forms of the flowerpots and blending the colours of the shadows.

FLOWERS

When painting flowers it is largely a matter of evaluating tones and not, as one might think, of looking first at colour. It is easy to be diverted by the sheer brilliance of colour in flowers and to lose sight of the tonal values completely. It is these tonal values which, as in everything, enable us to describe form: the delicate roundness of a bloom, the way a petal curls, the veins and ridges on a leaf. Try not to get bogged down in describing each individual flower in exact botanical detail; think of your vase of flowers as a single form in itself and establish the large areas of colour and tone first. Starting from this base, you can then build up gradually to the more complex and descriptive details. As the great French Impressionist, Renior, said: "When I am painting flowers, I establish the tones, I study the values carefully, without worrying about losing the picture..."

LIGHTING

Central to the question of tonal values, of course, is lighting. Look carefully at your subject to see how the shadows and highlights created by your light source fall. Are the patterns of light and shade clear and strong? Or does your subject appear to be a confused jumble of indeterminate tones? Keep altering the position of the flowers in relation to the light source until you are happy that the tones and colours are clearly described.

Strong, direct light accentuates form by creating sharp contrasts of tone, so the simplest approach to painting flowers indoors is to use a spotlight. If you wish to paint by natural light, choose north light if possible, because it is the most consistent.

Of course, you may wish to use a particular form of lighting to emphasize a particular mood. A diffused, softer light shows the subject as a series of subtle tones. Backlighting can create stunning effects, in which some parts of the arrangement catch the light while others are bathed in shadow. With front lighting you will get very little shadow, but the colours and patterns of the flowers are shown to the best advantage.

BACKGROUNDS

The background should be considered right from the start, since its tone and colour influence everything in the painting. For example, is your painting to be high key or low key? Your decision will bear on whether your background will be dark or light.

There is much to be said for setting light flowers against a

◁◁**Irises 1** by Lucy Willis, watercolour,
15in×9in (37cm×22cm).
Strong sunshine illuminates the flowers
from behind, making the petals appear
translucent and their colours intense.

◁**Irises II** by Lucy Willis, watercolour,
15in×11in (37cm×27cm).
Here the light comes from the side,
creating quite a different effect.

△**Spode Jug with Late Summer
Flowers** by Pamela Kay, oil on board,
13in×14in (32cm×35cm).
Placed against a rich, dark background,
the blue and white flowers stand
out strongly, their cool colours
complemented by the warm, reddish-
browns. Notice the range of greys that
have gone into the shadows on the white
flowers, and how these help to describe
their form.

dark background, and vice versa, since this provides a contrasting tone against which the flowers will stand out, thus helping to create the illusion of form and space. This is easy enough in oils as you can paint the flowers over the background, or work the background in at various stages as you paint. Watercolourists, however, may have to cope with the possibility of laying down a background wash around the intricate shapes of the flowers, without allowing the paint to dry in the process. Work steadily from one side of your picture to the other, pushing the edge of the paint continuously, so that it does not dry to a hard line while you are busy following the contours of the flowers.

THE PAINTING PROCESS

If you like, you can start by making a light outline sketch of the flowers on your paper or canvas, though this is not necessary if you have enough confidence to plunge straight in. A drawing can, in fact, be a distraction because it can trap you into using too many fussy details, and tempt you into thinking that painting is simply a matter of colouring in shapes.

It does not really matter where you begin a painting, but many artists prefer to start by blocking in the predominant colour. There may be a particular pinky-grey, for instance, which occurs in the background as well as the flowers.

Look next for the darkest areas – this may include the background – and mix the appropriate colours. Apply these darks wherever they may occur throughout the composition. Having done this, look for any patches of intense colour. This may be the yellow centre of a daisy, the luminous petal of an iris with light shining through it, or the bright green of a leaf. Because you have already established the darks and neutrals, these colours will stand out in vibrant contrast. The lightest patches of leaf, flower and vase can now be established. In oil or acrylic this is a matter of mixing very pale colours by adding just a small amount of pure colour to white; in watercolour, the very palest wash of colour can be applied and slivers of white paper can be left bare to indicate the brightest highlights. (Some watercolourists, of course, prefer to work the other way around – from light to dark.)

Remember that shadows on flowers, as on anything else, are seldom just a darker version of the local colour; they will be absorbing and reflecting colour and light from all around. The shadows on white and pale flowers are particularly beautiful; rather than mix a flat, dull grey, look for the subtle hints of warm and cool colour – blues, greens, pinks and yellows – that are present in the shadow areas.

▷**Honeysuckle and Flowering Bramble in a Jar** by Jacqueline Rizvi, watercolour and body colour, 11in×8½in (28cm×21cm).
In this closely-observed flower study the artist has paid as much attention to tones as to colour and texture. In so doing, she has created an impression of light and space as well as accurately describing each flower and leaf.

◁**Sun Seekers** by Frank Nofer, watercolour, 17in×13in (42cm×33cm).
Painting flowers has as much to do with seeing tones as it has to do with colour. Although these tulips are vividly colourful, it is the arrangement of light and dark tones which, ultimately, makes this painting so successful.

PORTRAITS AND FIGURES

You will be aware, by now, that the direction and intensity of the light can radically affect the mood and emphasis of a picture. So when setting up your portrait, think carefully about the lighting and do not hurry the process of posing your sitter. Decide what mood you want to convey and the characteristics that you wish to emphasize in your sitter. For example, if your subject is a girl reading a book, you may wish to convey a mood of quiet contemplation. You could place the girl in a comfortable chair next to a table lamp which sheds soft, warm light onto her face.

Look at the portraits in this chapter, and at any others you can find in art galleries. Try to work out how the lighting has been arranged, and how this determines the effect the artist has achieved. If the light comes from a single source, such as a lamp or a candle, the shadows and background will be relatively dark because there is very little diffusion of light to illuminate the shadows. This kind of lighting can give to a portrait a strong, sombre, even mysterious air. If, on the other hand, the portrait has been painted in a room with large windows, or in a studio with good north light, the figure, the face and the background will be lit with a diffuse, all-pervading light. The tonal contrasts will be gentler and more subtle, and the overall effect will be bright, open and clear.

PAINTING THE PORTRAIT

Painting the human form and face has been regarded by many as the apogee and pinnacle of art. However, there is no need to treat this subject in any way differently from any other; the skills of observation required are exactly the same. This book is about painting light, but it will not go amiss here to mention the importance of drawing. Drawing is the key to an understanding of proportion, and achieving a 'likeness' in a portrait is largely a matter of judging proportions accurately. Though a portrait should always be more than

just a mechanical representation of the sitter, scant likeness can be achieved without studying the spaces between and around the features. This is the framework on which the artist builds with tone and colour.

By drawing, however, I do not mean that you should make a precise preliminary outline of your subject, since there is a danger that this will constrict you and inhibit the flow of your painting. It's far better to indicate outlines and features with loose strokes which simply serve as a guide for the completed portrait – not a border that cannot be crossed.

It's the same when you begin painting; try to ignore individual features and details in the early stages and look at your subject as a series of large, interconnected masses. The face does not consist simply of a flat surface from which a few features protrude, but of a complex mass of angles and planes which receive and reflect light in a way which describes how

these planes come forward or recede. When the face is lit in such a way as to emphasize its structure – by three-quarter light or side light, for instance – each plane and facet can be seen as a change in colour or tone. It makes sense, therefore, to block in with appropriate hues the main lights, half-tones and shadows, working quickly over the face, the background and the clothing so that everything flows together with a unified continuity. Use the biggest brushes that you can comfortably control, and work with broad, sweeping strokes that follow the general form.

As you proceed, your description will become fuller. Build up the planes with patches of paint wherever you see a change in tone or colour, while keeping a clear idea of the main tonal relationships; do not overdo the darks on the light side of the face and keep the changes in tone subtle in the shadows. Look hard for warm and contrastingly cool colours in the shadows;

▷ **Portrait of My Father** by D J Curtis, oil on canvas, 18in×14in (45cm×35cm). Soft, diffuse light enters the picture from the left, creating a rim of light on the contours of the face. Sensitive brushwork and cool, high-key colours contribute to the contemplative mood of the painting.

◁ **Sue and Stephen, Not Talking** (detail) by Lucy Willis, oil on canvas, 22in×16in (55cm×40cm).
The room was lit by a window behind me which threw plenty of light onto my canvas but very little onto the figures. I set up a desk lamp on the floor, to the left, and directed it toward the group. This produced a warm, upward light, almost suggestive of firelight, and intensified the blue in the shadows, which absorbed the cool daylight from the window.

△**Pocono Angler** by Frank Nofer,
watercolour, 15in×11in (37cm×27cm).
The numerous planes and facets which
describe the structure of the face are
accentuated here by strong, direct light
falling from the right. Frank Nofer uses
strong colours and vigorous brushwork
to capture the angler's craggy features.

▷**White Fur** by Sally Strand, pastel,
57in×36½in (145cm×93cm).
The way in which light strikes the figure
is all-important when painting portraits.
Here, strong side light has been used to
great effect, lending dynamism to an
impressively statuesque pose.

you may find that you need to mix deep reds and oranges to bring out the richness within the darker areas.

Once the forms are well established you can begin to look more closely at individual features and the way in which light picks them out: how it reflects in the eyes, creates highlights on the nose and describes the modelling of the lips. It is important to do this towards the end of the painting and not get involved in such detail too early on; while a carefully observed rendering of the eyes, nose and mouth is essential to a good portrait, no amount of descriptive skill can make up for an underlying weakness in the overall tonal structure.

MIXING FLESH TONES

People often have difficulty in knowing which colours to mix for painting flesh tones. It can take quite some time to come to terms with the idea of using a varied range of colours to describe what appears, at first glance, to be simply pink, or black, or brown. In fact, the skin is a highly reflective surface

and its colour is influenced by the prevailing light and by colours reflected onto it from nearby objects. So, for example, the skin tones of a person sitting next to a table lamp, or a glowing fire, will contain a high proportion of warm pinks, reds and ochres; that same person's skin, viewed outdoors under the shade of a tree, will appear generally cooler in colour with hints of blue and green reflected from the sky and the foliage. It takes a little courage, when painting skin, to start applying green or blue! But it is precisely these hints of reflected colour that give the skin its translucent appearance.

The basic colour of skin can be mixed from the three primary colours: red, yellow and blue. Thus a good starter palette might consist of colours such as cadmium red light, cadmium yellow and cobalt blue, plus white (except in watercolour) and possibly some of the earth colours such as burnt sienna and burnt umber. These are only suggestions, of course, and every artist will develop his or her own favourite 'formula'. The point is to be aware that skin consists, not just of a single, flat colour, but of many shifting layers of colour. With your basic palette you can mix an infinite variety of flesh tints, adding more whites, reds and yellows for the warm, light-struck areas and more blues and greens in the shadow areas. In those areas of the skin that reflect strong colour from the immediate surroundings, you may wish to apply pure colour from the tube; small flecks of red, yellow or blue can enrich a passage and bring it to life. Similarly, you could try a mixture of just two pure colours, giving you a range of oranges, greens and violets.

It is important to pay particular attention to the colour of the shadows on skin. If you simply darken the colour with black, there is a real danger that you will make your sitter appear terribly ill. Do not, then, simply cool the colour of your shadows, but look also for warm glowing reds and oranges reflected in them. Most painters are, to begin with, very tentative in their use of strong colours on flesh, but it is amazing what boldness you can get away with – and how enormously your painting can be enriched for it.

▷**Verlayne Reclining** by Arthur
Maderson, oil on canvas, 32in×44in
(81cm×112cm).
Close harmonies of colour and value
contribute to the gentle tenderness of this
lovely painting. Maderson is less
concerned with describing the interior
and figure *per se* than with describing
the way sunlight fills the room and
warms everything it touches. The
reclining nude is the focal point of
the picture, yet at the same time the
figure seems almost to melt into the
surroundings.

△**Linda** by Lucy Willis, oil on canvas, 24in×20in (60cm×50cm).
If you half-close your eyes you will see that the tones in the figure are very close to those in the surrounding blue shadows; apart from the pale rims of light it is mainly the difference in colour temperature that distinguishes the two. This effect, created by backlighting, is what fascinated me initially about the pose, and I tried to keep the idea in the front of my mind throughout the painting.

Demonstration: Peggy Mathias

A few days before starting my portrait of Peggy I prepared the canvas with an overall purplish-grey. This consisted of the colour created when I mixed all the spare paint on my palette together at the end of a day's painting, and it gave me a good basic undertone on which to work. I spent some time studying the pose and deciding where to place the figure on the canvas in relation to the surrounding areas of space. I eventually decided Peggy should sit so that I had a three-quarter view of her face, strongly lit by the light from a north-facing window.

1

2

3

1 I begin by sketching in the outline of the figure with dark grey paint thinly diluted with turpentine. Using a palette knife, I mix three different colours: a reddish-brown for the flesh, a blue-grey for the sweater, and a very pale cream for the lightest parts of the face. These are applied with brisk, vigorous brushstrokes using a separate brush for each colour. I block in the skirt and the darkest part of the background with ivory black.

2 I continue to block in the main areas, adding pure cobalt blue behind the head and white on the cup. Then I start to pay more attention to the planes of the face and head, introducing both warm and cool colours wherever I come across them. Most of the flesh tints are made up of just three colours and white, mixed in various proportions: cadmium red, cadmium yellow and a little alizarin violet or cobalt blue. In addition to a handful of hogshair brushes, I begin to use a few softer, more pointed sables so that I have greater control as I apply smaller and smaller patches of colour.

3 Now I begin developing the forms of the body and legs, paying special attention to tonal values. I draw the hands a little more informatively, and begin to work more closely on the features of the face. I frequently stand back and look at the painting from a distance to see how it is shaping up. I am aware that if I get too involved in the details of the face I may unbalance the composition as a whole.

137

4

5

4 I realize that the composition requires a little more interest on the right, so I move the biscuits and sugar bowl into the picture. The background and table are toned down in relation to these white objects. I use some of these mixed greys to introduce more colour into the sweater.

5 Titian said that if you left a painting face-to-the-wall for long enough, it would paint itself, and this, I very often find, is true. A painting on which you have been working intensively is impossible to see clearly until some time has passed; only then do the faults become clear. So, after an interval of two days I look again at the portrait and resolve to make a few changes. The limited palette of black, grey, and flesh colours, which I had originally conceived, does not entirely satisfy me, so I ask Peggy to put on a bright red skirt and a patterned Thai jacket. I quickly obliterate the legs with a vivid combination of crimson lake and cadmium red, and use the same colour for some of the stripes on the jacket. I also decide to break the stiffness of the pose, not by moving the body, but by lifting the eyes so that they look out of the window and reflect more light.

6 In the final stage I enjoy myself painting the patterns on the clothing, taking care not to put too much light into the shadow areas, which would disrupt the impression of strong light falling on the upper part of the figure. I make the coloured stripes on the sleeves appear to curve around the arms by grading their tone and intensity. Finally, I neutralize the blues and reds in the background so that they recede, and I work over the rectangle on the right with broken strokes of warm and cool greys to activate the surface of the painting.

▷ **Peggy Mathias** by Lucy Willis, oil on canvas, 24in×20in (60cm×50cm).

6

INDEX

CREDITS